THE WINDMILLS OF SUSSEX

To my wife Mary
and daughters
Lesley, Katherine and Sally

THE WINDMILLS OF SUSSEX

MARTIN BRUNNARIUS

PHILLIMORE

1979

Published by
PHILLIMORE & CO. LTD.
London and Chichester

Head Office: Shopwyke Hall,
Chichester, Sussex, England

ISBN 0 85033 345 8

Printed in Great Britain by
UNWIN BROTHERS LIMITED
at the Gresham Press, Old Woking, Surrey

and bound by
THE NEWDIGATE PRESS LTD.
at Book House, Dorking, Surrey

CONTENTS

LIST OF FIGURES

LIST OF PLATES

(between pages 52 and 53)

ACKNOWLEDGEMENTS

I would like to make the following acknowledgements: to the dedicated work of the late H. E. S. Simmons, from whose collection I have drawn (with the kind permission of Mrs. E. Simmons).

To those Sussex mill folk who have given extensive help and friendship, including Mr. Sidney Ashdown, the last miller at Cross in Hand; Mr. Gilbert Catt of Hamlin's Mill, Hailsham; Mr. A. D. Dallaway, owner of Punnett's Town Cherry Clack Mill; Mr. Frank Gregory, a unique Sussex windmillian; Mr. Ron Hall, owner of Stone Cross Mill; Mr. Clifford Hoad of Bexhill Down Mill; Mr. C. J. Newnham, owner of Cross in Hand Mill; Mrs. Bertha Terry, daughter of Albert Ovenden, the last miller at Polegate.

To Mr. Rex Wailes, O.B.E., F.S.A., F.I.Mech.E., of the S.P.A.B., for constructive criticism and advice and for contributing the Foreword.

To repositories and staff in the county, including: Brighton Reference Library; Hastings Reference Library, Mr. Purdey and Mr. Powys; Hove Central Library, Judith Dale; Worthing Reference Library, Mr. R. Elleray; East Sussex Record Office, Mr. A. Dibben and Mr. S. Hewson; West Sussex Record Office; Goodwood Estate archivist and Mr. Harmer; Borough of Hove, Mr. Ray; Wealden District Council, Mrs. P. Davis and Mr. F. Child; Newhaven and Seaford Historical Society, Mr. V. H. Bourne; Sussex Archaeological Society, Fiona Marsden; Weald and Downland Open Air Museum, Mr. C. H. Zeuner; Winchelsea Town Museum; Worthing Town Museum; Eastbourne and District Preservation Society; Nutley Preservation Society, and Rottingdean Preservation Society.

To those who have allowed me to draw from their own research, including Mr. J. Mainwaring Baines of Hastings; Mr. H. R. Hawksley; Mr. H. C. P. Smail of Worthing; Mr. L. Stevens' extensive research work on Eastbourne's horizontal mills.

To the owners of windmills and remains in the county, who responded encouragingly to my quest for additional material, and gave permission to photograph their mills and homes in particular: Mrs. V. Aslett, Kingston; Mr. G. Beecher, Chapmans Town; Boardmans Garage, Ore; Mr. D. W. Bourne, Golden Cross; Mrs. Bull, Northiam; Mr. and Mrs. R. Buckmaster, Hurstmonceux; Mr. J. Cornford, Punnett's Town; Mr. A. D. Dallaway, Punnett's Town; Mr. C. N. Darby, Earnley; Air Commodore J. F. Davis, Washington; Mr. R. J. de Little, Keymer; Mr. C. J. Dowd, Patcham; Mrs. H. Evans, Wisborough Green; Mrs. D. K. Freeland, Ore; Glyndebourne Estate Trust; Mrs. J. Greenall, Baldslow; Mr. R. Hall, Stone

Cross; Mr. Hawkins Byass, Keymer; Mr. C. J. Hoad, Bexhill Down; Mr. A. J. Holcombe, Crowborough; Mr. J. Leberne, Ditchling; Mr. and Mrs. Lee, Lewes; Mr. G. Llewellyn, Dicker; the late Henry Longhurst, Clayton Mills; Manager P.O. Pension Fund Home, Climping; Mr. Martyr, Harebeating; Mr. W. Mason, Dell Quay; Mr. G. L. McNally, Battle; Mr. P. Merricks, Icklesham; Mr. J. Michell, Arundel; Mr. J. R. Miller, Hunston; Mr. F. Mora, Chailey Heritage; Mrs. D. Mutter, Chailey South; Mr. C. J. Newnham, Cross in Hand; Mr. Overton, Riseden; Mrs. N. Perigoe, Northiam; Mr. and Mrs. C. Redman, East Wittering; Mr. R. C. Reynolds, Barnham; Mrs. M. Robson, Hastings; Mrs. H. B. Roscow, West Chiltington; Mr. G. Smythe, Lewes; Mr. E. V. Tant, Walberton; Mr. S. Tubb, Seaford; Mr. Weake, Ashurst; Mr. A. F. Webb, Rye; Mr. R. W. H. Wicking, Staplecross; Mrs. H. Woodward, Malling; Mrs. J. Yates, Fishbourne.

To the following sources of photographs: 122 are from the author's collection, the remainder: Mr. S. Ashdown, 70, 93, 96, 101, 111, 116, 117, 122, 129, 153, 154, 160, 162, 197; Mr. F. W. Gregory, 17, 29, 33, 86, 108, 130, 158, 166, 170, 194, 197; Edward Reeves collection, 8, 53, 56, 82, 104, 109, 135, 184; Gurney Wilson collection, 48, 56, 66, 71, 85, 105, 114, 125, 136; Mr. G. A. Catt, 62, 75, 81, 195, 196; Mr. Rex Wailes, E. L. Burne collection, 102, 128, 200, 201; J. M. Baines collection, 142, 143, 146; Brighton Reference Library, 68, 77, 148; The Harmer family, 41; Mr. F. Child and Wealden District Council, 169; H. W. Baylis & Son Ltd., The Arcade, Littlehampton, 121; S. Ashdown/J. Russell, 150; Mr. A. D. Dallaway, 59; Mrs. J. Greenall, 117; Mr. R. Hall, 89; Mr. P. Hill, 79, 91; Mr. E. Hole, 36, 179; Mr. R. J. de Little, 191; Mr. J. K. Major, 58, 199; Newhaven and Seaford Historical Society, 193; Mr. C. J. Newnham, 23; H. E. S. Simmons collection, 11, 37; Worthing Reference Library, 164, and for specialist printing services, to Park Cameras (R. A. Atkins), Burgess Hill and Dana Sykes, Burgess Hill.

Last, and most important of all, to the forebearance and patience of my family, in particular my wife Mary, who spent countless hours converting my longhand notes into a readable typescript.

FOREWORD

THE AUTHOR of this excellent and most attractive book has a thorough knowledge of the Sussex windmills and imparts it clearly. The fact that he has been able to draw upon the researches of the late H. E. S. Simmons is a guarantee of its accuracy. The illustrations greatly enhance it and, best of all, is the attention given to the millwrights who built and repaired the mills. I commend this book to all who are interested in windmills in general, and Sussex windmills in particular.

REX WAILES

President, Wind and Watermill Section,
Society for the Protection of Ancient Buildings

CHAPTER ONE

INTRODUCTION

SUSSEX WINDMILLS, figuring as part of English history since early times, stand today as memorial to industrial heritage lost.

This statement reflects perhaps an overall nostalgia which is shared by many who are drawn to the remaining few, some of which have an air of sadness about them, others more joyful.

The county has boasted windmills since their introduction in England, and indeed may well boast the earliest positive reference to one in the country. Joseph Cooper, F.S.A., made a contribution to the 1879 Sussex Archaeological Collection[1] concerning 'The Hundred of Swanborough', and in this, an entry from the 'Testa de Nevill' reads:

> 'Hugo de Plaiz gave to the monks of Lewes, the windmill in his manor of Iford, for the health of the soul of his father'

and is dated 1155.

The mill in question may have stood on Iford Hill immediately above Swanborough Manor and in all probability was supported by a sunk post, evidence of which may yet be found.

A later and well-known reference[2] dates from 1264. During May of that year, King Henry III and his brother Richard (King of the Romans) came to Lewes to seek support for the Royalist cause against his brother-in-law, Simon de Montfort, and on the 14th of the month, the two opposing forces met in battle on the Downs to the west of the town. De Montfort and the barons routed King Harry and finally captured him at the Priory below the castle. Meanwhile, Richard had taken refuge in a post-mill which stood on the slope of the Downs about a quarter of a mile west of the castle. The barons, discovering Richard, had him brought out to cries of 'come down thou bad miller' and similar taunts.

This is the first note that we have of a Lewes windmill. The demise of the last to stand in the town is recorded in Chapter Six, spanning 650 years or so.

In the county as a whole, of the hundreds that stood and fell over the intervening centuries, those extant today represent in visible form, the final 150 years or so of serious consideration to the art of grinding grain and other edible seeds by windpower.

In earlier times, windmills were plentiful near the coast and on the Downs. Here, the warm south-west winds prevail and crude post-mills were built to work with this. Inland, the Weald was often served by water-driven mills. However, during the growth of the Sussex iron industry, the many inland rivers and streams became fully utilised for water-driven hammer-mills at the forges. These and fulling-mills for the woollen industry left little room for new flour mills.

Thus the windmill began to appear on all of the high and favoured sites of the Weald, and with the coming of full agriculture, the whole county was was attractively dotted with hundreds of wind- and water-mills.

The wind, always a tempting source of power, unlike the flow of water in streams, was never easy to harness satisfactorily. Even the most up-to-date mills, like those at Stone Cross and Shipley, had weeks at a time when there was not enough wind to be useful.

During the early history of the windmill generally, the difficult task of converting the wind's horizontal force into a vertical rotation of enough power to grind was overcome. Cloth sails were rigged to spoked frames supported by a horizontal shaft. These were allowed to catch the wind much as a ship's sail does, thus giving the required rotating motion.

The shaft here would have carried a primitive trundle wheel and the drive taken from this direct to the top or runner stone. This was, of course, a derivation of the donkey-, ox- or horse-mill which, as the names suggest, was driven by a beast of burden endlessly treading the same circular path. The animal was harnessed to an arm and cog ring which turned the runner stone via up upright shaft. I believe that the round-houses used for this can still be traced today in some cases.

The vertical sail windmill became firmly established as a method of grinding and various methods of machinery mounting were adopted. The construction of the body varied from place to place. Wood was usually the cheapest and quickest to use, and so the open trestle post-mill, being entirely of wood, evolved in its early form, especially in Sussex, where the material was plentiful. The whole of the body was made to turn about a central post. The open trestle allowed the wind to pass all round the mill. Sails or sweeps, as they are called in Sussex, were made of frames set at the required angle to be driven by the wind when carrying cloth. The body supported the windshaft and also the stones, which were crudely dressed more to crush the grain than cut it. The whole would be pulled round into the wind by manpower using a tailpole. So here we have the basic early post-mill (Plate 1) mounted on a simple trestle to give the required stability. The tower-mill may be the earliest form of windmill, used a lot in European countries in early times. This type came later in Sussex as did the smock-mill, which, strictly speaking is a wooden tower-mill.

Later, as the mill became more interesting to the engineer of the times, this engine of the wind was further refined, with innovations such as automatic winding which could only be successfully applied to well-built and balanced mills. The sweeps were more scientifically built as carefully-angled frames upon which canvas could be placed and furled according to the strength of the wind. Shuttered sweeps followed with pivoted slats spring-loaded against the wind. This partially automated the mill, Cubitt's patent coming later to give the last major refinement. As grinding and mills became more sophisticated, two, and later, three pairs of stones were fitted. These were very much more finely dressed to give the refinement and texture of meal demanded. Specialised equipment evolved from the primitive sieves used by predecessors, becoming bolters and dressers for grading the meal, smutters and blowers for cleaning the grain.

Construction also progressed through the years. The post-mill structure we see today, with its massive oak timbers so well fitted and balanced, supported by likewise built oak or pine trestles, is the result of centuries of skill handed down, most of it traditional rule-of-thumb like all country crafts. The smock- and tower-mills developed similarly. Nothing we find today is truly standard. Each mill is unique in its own right, a most gratifying gift of past ages. Today in Sussex all three main types of windmill survive with their various merits.

Being primarily concerned with the mills standing in this county, we shall consider the relevant details and points relating to Sussex examples. For a broader and informative outlook countrywide, so to speak, a bibliography is to be found at the end of the book which will adequately reward the reader in the further pursuit of this interest.

The various terms and phrases used in mill terminology are mostly common to the whole of England. These come to us giving a glimpse of the countryman's English. Using words to describe the special components and features by relating to everyday things of the time. Here are several examples:

Crown Tree: the huge horizontal baulk which carries a post-mill ('crowning the post);

Side Girts: the 'great' side timbers supporting the post-mill body at the ends of the crown tree;

Damsel: part of the feed mechanism which chatters endlessly whilst working;

Spider: an iron coupling assembly. This appears to be sitting at the centre of the sweeps.

The glossary and index defines the more obscure terms used here and later, and will help the reader.

After this brief introduction, the following chapters deal in more detail with the basic construction of Sussex mills, the mills themselves, their environment, and the people associated with the scene past and present.

CHAPTER TWO

WINDMILL TYPES

FIRSTLY, it should be said that Sussex boasted most of the hybrid variations such as Hollow post-mills, Horizontal mills, and large wind-wheel structures used as prime movers. Apart from the reconstruction of a hollow-post pumping-mill, almost nothing remains of these unusual windmills. However, the reader will find photographs and notes of them in the sections concerning Remains and Bygone Miscellania.

Today there are three main types of windmill to be found in the county. They are:

i. The post-mill.
ii. The smock-mill.
iii. The tower-mill.

and we now look at the basic features of their construction.

(i) **The Post-Mill** (Fig. I)

The main structure of this type is built entirely of wood. Starting from the bottom and working up, inside what is called the roundhouse, built of brick, stone or wood, we find that the whole is supported by a huge timber frame known as the trestle. This structure (Plate 1) is the base of support for the mill, composed of the upright centre post, supported by two interlocked triangular frames. The base of each of these frames is a timber baulk, called a cross tree, about 20ft. long and resting horizontally on brick piers. The two cross trees pass one over the other, specially notched and morticed at each end (Plate 3) to support the quarter bars. These, morticed into the post just below the body itself (Plates 95 and 110), take the total weight of the mill. The post is usually cut from oak and is basically the trunk of a tree. Starting at its square base, this post has horns (Plate 2) which locate it over the two cross trees, keeping the mill upright. At this point, it can be almost 3ft. square. From here it rises up the point where the quarter bars join it. Above this, the post is always round, passing through a wear in the spout floor of the mill to the main pivot bearing in the crown tree above. Here the later centre post is usually fitted with a flanged iron pin or pintle, this is fitted on to the top of the post. The crown tree has a complementary iron thrust-bearing in the form of a large cast bracket (Plate 94). Earlier mills relied on a combined location and thrust-bearing formed by a direct wood-to-wood pintle and socket joint between the two. These suffered with excessive wear and friction and were a source of worry to the miller.

At the stone floor we find the crown tree itself spanning the body left to right. On each end of this rest the side girts, these huge baulks, sometimes known in Sussex as summer trees (from bressumer, I believe), run from the back to the front of the mill, and carry the rest of the framing. The whole structure is indirectly hung from or rests on these.

When building up the framing of the mill, the millwright had to arrange that its centre of gravity was somewhat behind the post bearing. This was counterbalanced by the combined weight of the sweeps and windshaft plus the stones and machinery acting forward of these bearings. A millwright, such as Medhurst of Lewes, would no doubt have allowed the mill a small amount of forward bias on its post to counter the force of the average working wind. A large hand-winded mill such as Hurstmonceux would need to be very well balanced indeed to keep the miller's effort to a minimum.

Four corner posts form the uprights for the body. The two at the breast or head of the mill support the main timber for the windshaft bearing; this is called the breast or weather beam. In the back or tail of the mill, the end of the windshaft is carried by the tail beam. Along the top of the sides are the top rails which carry the roof rafters and joists. At the bottom of the body, the spout floor timbers form a bearing or 'wear' around the post, coupling with the crown tree bearing to keep the body vertical on the post. Two large front-to-back timbers called sheers also pass under the body here to stabilise the whole structure. On inspection, we find many smaller timbers serving as struts and bracing to share the load imposed by the sweeps, windshaft, driving gear, and the weight of the stones themselves. Nearly all the large timbers are of oak or pitch pine.

The outside of the body is always weatherboarded, usually double lapped, with small windows or claps in each side. The breast and roof are weathered to improve the wind-flow and prevent the ingress of rain. Later mills were often over-clad with iron sheeting, especially as the weatherboarding deteriorated. Originally, copper or zinc was used for this purpose, especially on the roofs. Whilst being cheaper, iron has the distinct disadvantage of pulling at the securing nails with changes of temperature. The main entry to the mill body is by a stable door at the back, gained by steps from the ground.

A windmill is, of course, designed to work facing into the wind. If the wind is allowed to get behind the sweeps, patents particularly, it could be forced to run backwards, even against the brake, and in extreme cases, the whole sweep and windshaft assembly would be torn from the bearings. To keep the mill into

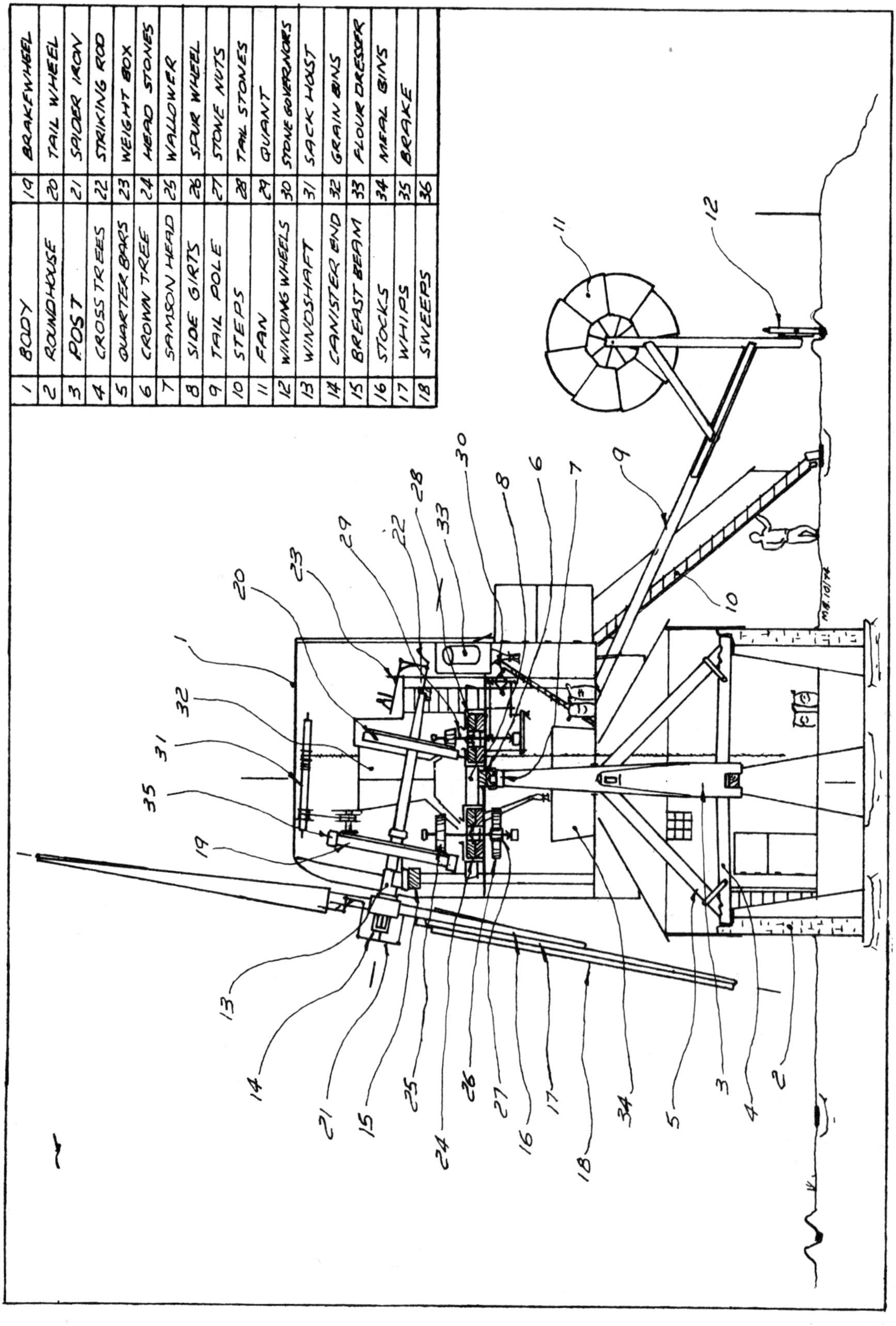
1 BODY
2 ROUNDHOUSE
3 POST
4 CROSS TREES
5 QUARTER BARS
6 CROWN TREE
7 SAMSON HEAD
8 SIDE GIRTS
9 TAIL POLE
10 STEPS
11 FAN
12 WINDING WHEELS
13 WINDSHAFT
14 CANISTER END
15 BREAST BEAM
16 STOCKS
17 WHIPS
18 SWEEPS
19 BRAKEWHEEL
20 TAIL WHEEL
21 SPIDER IRON
22 STRIKING ROD
23 WEIGHT BOX
24 HEAD STONES
25 WALLOWER
26 SPUR WHEEL
27 STONE NUTS
28 TAIL STONES
29 QUANT
30 STONE GOVERNORS
31 SACK HOIST
32 GRAIN BINS
33 FLOUR DRESSER
34 MEAL BINS
35 BRAKE
36
M.B. 10/74

wind and prevent it becoming tail-winded, the whole body is moved around the post. The procedure is called winding* the mill. In post-mills, winding is via a tailpole or fantackle. The method varies now, but originally this was always effected by hand. The steps can be raised from the ground by a lever known as a talthur, thus allowing the miller to push the mill round. At Keymer post-mill, a yoke was provided to enable the miller to use his weight and strength to give better purchase here. When the required position is reached, the talthur is released so that steps, in effect, peg the mill to the spot. Nutley post-mill is a perfect example of a tailpole and talthur mill. Later came the tailpole mounted fan. Here, the fan's action is transmitted via bevel and spur gears and a worm drive to the winding wheels which drive the mill round until the fan is out of the wind, i.e., when the full area of the sweeps is presented to the wind (see page 22). Cross in Hand and Argos Hill post-mills still have this type of tailpole fan-tackle. These are almost unique to Sussex. When this method is employed, the steps are supported by a pair of truck wheels. The mill is thus free to be winded by the fan, but unable to move otherwise.

Another method of winding in Sussex may be seen at Icklesham post-mill. Here there was no tailpole, the mill being brought into the wind by winding wheels at the base of the steps. The fan is mounted on the top of the mill body in a stage. Long interconnected drive rods communicating with steps below the tail of the mill.

The sweeps, which may be canvas rigged or shuttered, are carried by the windshaft, this being of wood or more commonly now, iron. One, or as is the rule in Sussex, two, main driving wheels are mounted on the windshaft and are called head and tail wheels respectively. The former is also known as the brake wheel because the huge band brake which holds the sweeps at rest, passes round its broad rim. These wheels are of wood or iron, invariably with wooden morticed cogs, and between 7ft. and 10ft. 6ins. diameter.

The stones are driven via pinion gears called 'stone nuts' attached to stone spindles. In a post-mill with head and tail wheels, the runner stones are driven direct from each wheel, this method of driving from above is known as 'overdrift'. A post-mill with a brake wheel only, usually drives two pairs of stones via a large gear known as a spur wheel. This is driven by a bevel gear or 'wallower' on the same shaft meshing with the brake wheel. The stones themselves are underdriven by stone nuts engaged with the spur wheel below. The two methods of driving from the windshaft are described as 'head and tail' and 'spur gear'. Cross in Hand post-mill had three pairs of stones when rebuilt. Two pairs in the head of the mill, spur gear driven, and a pair in the tail, overdriven from the tail wheel.

The part of the mill body where all this is found is called the stone floor, containing all the stones in their casings or 'tuns', the brake and tail wheels, and, of course, the windshaft. A huge lever passes along the body here (on the right-hand side in an anti-clockwise mill) often just above the steps, from the spout floor below. This is the brake lever (Plate 20). Very heavy, it has a fulcrum

*the 'i' in winding is pronounced as in win.

close to the brake wheel itself and can exert the full braking leverage at the release of the brake catch. When starting the mill, the brake rope is pulled, thus lifting the lever until the catch engages with a pin, this releases the brake wheel from the grasp or 'gripe' of the wooden band encircling its rim. The brake rope hangs down through the body of a post-mill. Above us here, the bin floor is situated. In small head-and-tail mills, this is not a true floor, but a part floor or even staging, enough to allow the miller access to the bins. Sacks of grain are raised to this position by a friction-driven sack hoist which employs what is known as a sack bollard and chain. Like the brake, this is brought into operation by a rope or lever system. Below the stone floor is the spout floor, named so because the meal spouts from the stones terminate here. The meal is either passed into a large bin to cool or direct into sacks, perhaps to be passed back to a dresser to be refined and graded.

In the spout floor we find above our heads the tentering beams and governing mechanism for the stones themselves, to be described in more detail, as are other parts of the mill's motion, a little later.

Taking a last look here at our post-mill, outside we can see today the early form as preserved for us at Nutley. This is a tailpole and talthur mill with an open trestle mounted on very shallow piers. The mill at Cross in Hand represents the Sussex post-mill in its final form being 'spur gear' with a tailwheel as well. The trestle here is mounted one storey up on brick piers and is itself enclosed in the first floor of its two-storey roundhouse; and of course there is a fine Sussex tailpole fantackle for winding the mill.

(ii) **The Smock-Mill** (Fig. II)

The main difference between this and the post-mill is that the working body of the mill which houses the bin, stone and spout floors is stationary. The cap alone rotates on a circular curb to bring the sweeps into the wind. This type derives its name from the likeness of the wooden tower to the smocks, universally worn by country folk, men, women and children alike, in times past.

The smock tower is octagonal in nearly all cases, the exception being West Blatchington, whose smaller body is hexagonal. Of the past Hunston smock-mill had 10 sides, and West Hill, Brighton had twelve. I believe that at one time in Sussex a post-mill with a tail fan was known as a smock-mill, why, I am not sure. As we have for the post-mill, we shall start at the bottom.

Here we find the 'base' as it is known. In Sussex, this is built of a number of materials, i.e., flints, sandstone, or brick, sometimes even a combination of all three. Between 16ft. and 24ft. across corners inside and one or two storeys high. Like the smock tower above, the base usually has sloped or battered sides. The total weight of the mill above is borne by these walls as a downward thrust. A much more stable approach this, to the problem

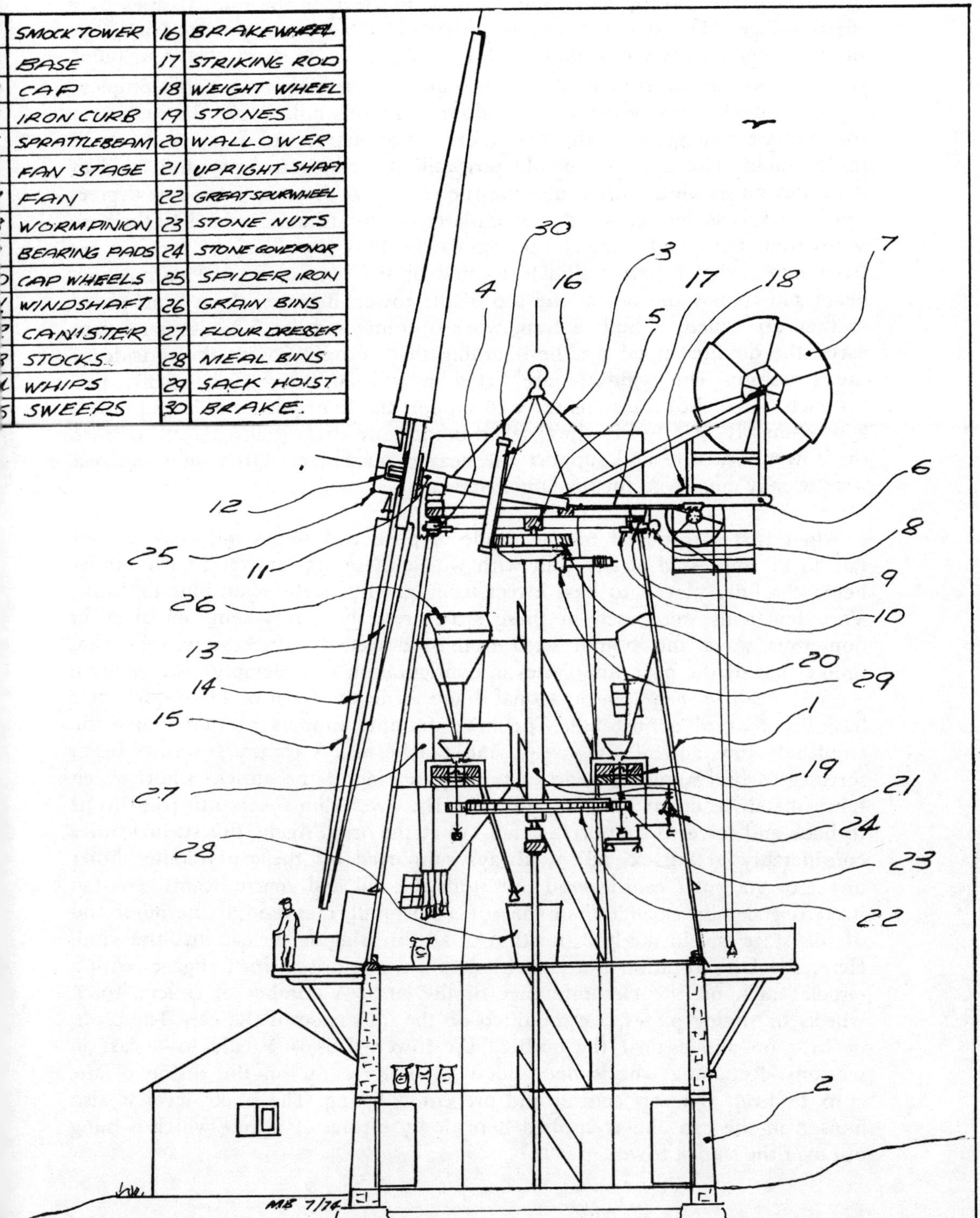

g. II. **Sectional drawing of a Sussex Smock Mill showing the main features (after Shipley).**

of keeping the sweeps aloft and housing the rest of the mill, it does have disadvantages. The smock tower here deprives the sweeps of more wind than does an open trestle post-mill, and I would guess is more costly to build. However, as can be seen in the smock-mills of the county, notably Shipley, there is much more room on the floors and the mill is not limited to a fore-and-aft arrangement, the tower being stationary, and as a result, much more stable. The body of an old post-mill is inclined to warp and wind as it rotates on its pintle, often altering the setting of the stones to some degree. The smock is seated on wooden sole plates which are usually held to the base with iron ties. Eight posts approximately 9ins. by 9ins. in section rise from here, with a batter similar to that of the base up to the curb which bears and retains the cap at the top of the tower. Between these 'cant posts' as they are called, is built a framework of joints and bracing, also serving to carry the double-lapped weatherboarding that entirely covers the outside of the structure. The windows are fitted in at each floor level, usually two for each. Here also, transoms serve to support the floor joists and huge timbers approximately 1ft. by 1ft. 6ins. called binders are strategically placed to brace the tower structure and support the heavier machinery. Pitch pine and oak were usually employed for the larger timbers.

When first introduced to the inside of post and smock-mills, we cannot fail to be impressed at the skill with which these are 'crafted', each timber being carefully placed to best effect using gravity as the main binding force. Very few nails were used in these structures, the joints being morticed in numerous ways and bound with iron or dowelled as necessary. The cap, unlike that of the post-mill, varies in configuration considerably. This may be a round beehive shape, an octagonal dome, or in the form of a hood, which is basically that of a post-mill. The main framing timbers of these carry the windshaft fore and aft on breast and tail beams. A central 'sprattle' beam serves as a brace and also carries the top bearing of the upright shaft, which takes the drive down into the mill (Fig. II). Two main sheers run from front to back and serve to support the fan tail at the back. Again, this staging varies considerably in Sussex, but is always supported by the cap framing either direct or via posts cantilevered out from the tail and centre beams. The fan itself is generally smaller than that of a post-mill, mounted at the outer end of this stage and in the best position of all for bringing the cap into the wind. Here, the fan's rotation is transmitted to a worm drive which engages with a circular rack on the circumference of the curb. A number of rollers, truck wheels or sliding plates, are mounted on the underside of the cap. These run or bear on and against the curb of the tower. This is usually iron, cast in sections. Retaining wheels bear sideways under a lip on the inside of the curb to keep the cap central and prevent it lifting. The brake lever is also housed in the cap and is applied here or by means of a rope which is hung out over the smock tower.

Originally, the smock-mill was winded by hand using a braced cap pole as seen on Dutch mills today, or via a wheel and chain. This drove the worm pinion direct.

Internally, the machinery of the smock mill is almost the same as that of the post-mill with one basic difference. To appreciate this variance, we must compare the two structures. The post-mill windshaft and its drives to the stones are carried around the centre post as a complete working unit in the body. In the smock-mill, the cap alone turns taking the sweeps and windshaft with it, the stones being housed in the stationary tower below. Thus the windshaft has a brake wheel only, travelling round the centre of the mill. This drives one large vertical shaft via a wallower wheel. The upright shaft, as it is known, has a top bearing in the sprattle beam of the cap and passes down into the mill, terminating below the bin floor as a great spur wheel and supported by a footstep brass. In this case the stones are overdriven by crutch poles or quants, three pairs of stones being most common in Sussex. In an underdrift arrangement the upright shaft passes through to a point just below the stone floor, allowing more freedom in the stone floor itself. The overdrift method in smock mills may be seen at Punnetts Town, and the underdrift at Shipley. The bins are in a proper floor, arranged above the stones, often with lids or 'claps' in the gear floor above. Here, a friction clutch on the underside of the wallower drives the sack hoist.

Descending the steps from the gear floor or cap, we pass the bins, then down through the stone floor to the spout floor, where, as with the post-mill, the spouts and meal bins are situated. The spouts are usually of wood and have two outlets to allow the miller to divert the meal from the bin to sacks for further dressing. The machinery for this is usually on the stone floor above and further spouts from this hang here also. As with the post-mill, we find above us here on the spout floor, the tentering gear for the stones and in underdrift mills, the great spur wheel and stone nuts. At this level, a stage, giving access to the weight chain, brake rope and sweeps is gained by two doors. Below are the upper base or loading floor (if two storey) and ground floor, used for storage of grain and meal.

(iii) **The Tower-Mill** (Fig. III)

This has developed from early beginnings as have the post- and smock-mills. In Sussex we find that its tower-mills are basically brick-built smock-mills and do not need further technical description with regard to machinery.

The brick tower, of course, is the most stable of all mill constructions and usually the most expensive to build. The most modern mills in England were of this type.

The main disadvantage of tower mills, apart from initial cost, is that the necessarily thick walls hold the damp, and unless the mill is properly faced, this is a problem. The outsides of Sussex tower-mills are either tiled or cement

1	TOWER	17	STRIKING ROD
2	ROUNDEL	18	ROCKING LEVER
3	CAP	19	STONES
4	IRON CURB	20	WALLOWER
5	SPRATTLE BEAM	21	UPRIGHT SHAFT
6	FAN STAGE	22	GREAT SPUR WHEEL
7	FAN	23	STONE NUTS
8	WORM PINION	24	STONE GOVERNORS
9	CAP ROLLERS	25	SPIDER
10	CAP WHEELS	26	GRAIN BINS
11	WINDSHAFT	27	DRESSER
12	CANISTER END	28	MEAL BINS
13	STOCKS	29	SACK HOIST
14	WHIPS	30	BRAKE
15	SWEEPS	31	QUANTS
16	BRAKEWHEEL	32	

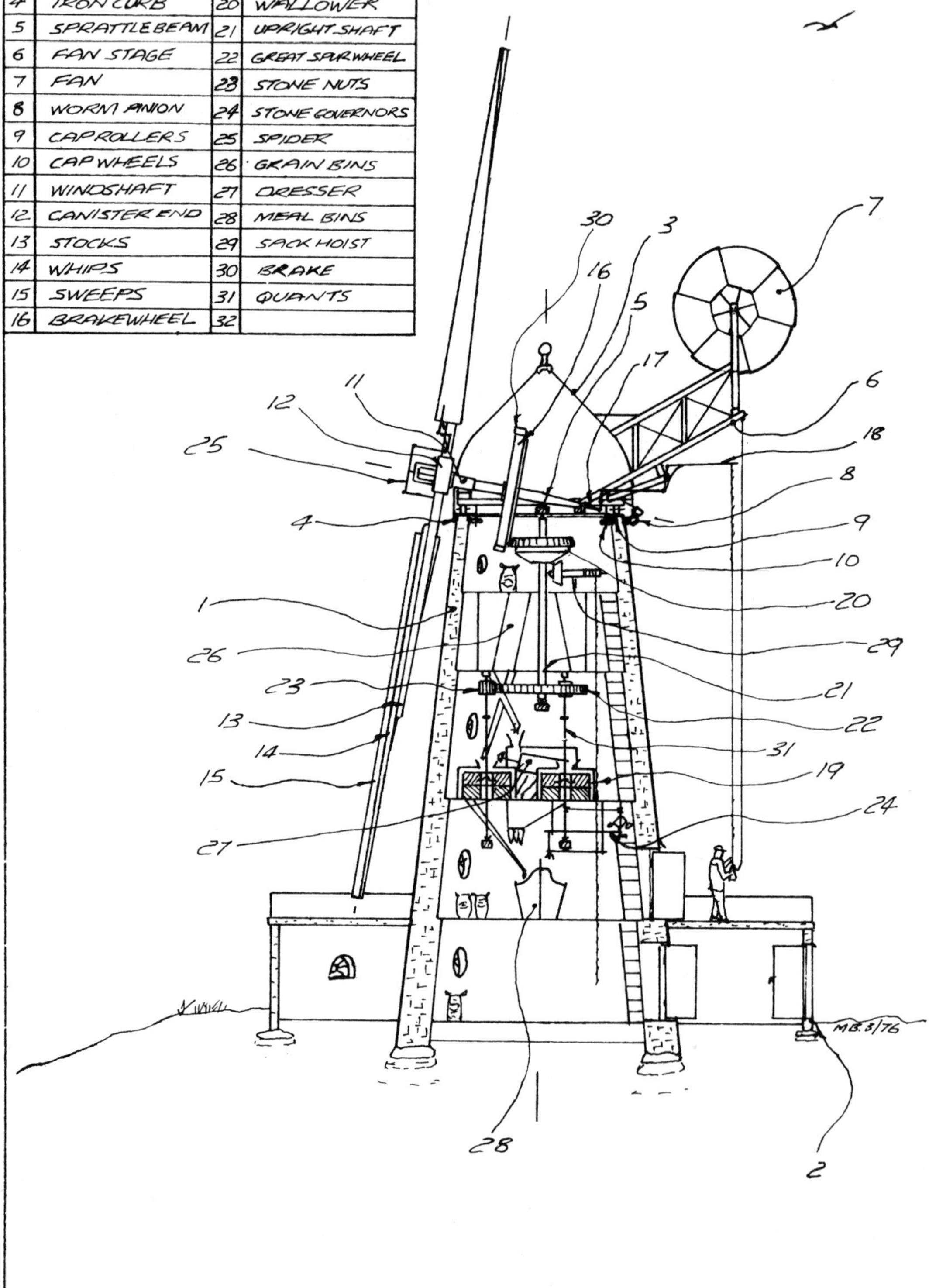

Fig. III. Sectional drawing of a Sussex Tower Mill showing the main features (after Stone Cross).

faced to prevent this. The tower-mill at Clayton which is very exposed, has metal cladding over cement. Often Sussex tower-mills were tarred with white sweeps and cap, giving a very masculine appearance when compared with the more fragile lines of a white post mill.

The main advantages are that the main structure does not suffer from the direct ingress of driving rain or with the infestations and plagues of the above-mentioned types, i.e., woodworm, the consumer of wooden mills, and such as wet and dry rot which convert the timbers into poisonous fungus. Nevertheless, the flooring and its timbers in the tower mill often suffered from damp at the edges. The usual method of supporting the joists is on the top of a reducing course in the brickwork, the walls at base level being perhaps six courses thick, reducing to two or three at the curb. The fan and cap mechanisms here are also similar to those of the smock-mill.

This concludes the brief description of the main structure of the Sussex windmill.

The next chapter deals more specifically with the moving parts of the mill.

CHAPTER THREE

THE MOTION OF THE MILL

(i) Sweeps and Windshafts

THE WIND'S ACTION has to be converted into a powerful rotation, this being achieved by setting the sweeps at an angle to the plane of rotation. The early cloth sweeps were set at a constant angle of weather, as it is called.

The cloth or common sail, originally known as a 'swift' in Sussex,[1] being the early form, was in use right up till the final decline of windmills around 60 years ago, when power milling came into its own. Latterly two 'commons' were used in conjunction with two 'spring' sweeps. We see a typical example of this type of sweep at Nutley (Plate 4). Built up in 18-inch squares forming a lattice framework, this may be between 25ft. and 30ft. long and 6ft. to 7ft. across, according to the size of the mill that they drove. The components of this are longitudinally: the hemlath which trails, and the leading edge or wind board at the front, with uplongs spaced between. Laterally, the frame is braced across with sailbars. A spar of pitch pine forms the support for the sweep. This is called a whip and the sailbars are morticed right through it. The angle at which the millwright chiselled out these joints determined the angle of weather which gives the vectored rotation to the sweep. Millwrights and engineers of the past found that to make best use of wind, the angle of weather should increase towards the inner end or heel of the sweep so that a well-made set of sweeps will have an attractive twist away from the windboard, sometimes known as the 'drift' in Sussex. This varies according to local tradition, but is typically in Sussex four degrees at the tip opening out to 22 degrees at heel. The portion of the sweep behind the whip, the broadest part that is, is called the 'driving' side and that part in front is called the 'leading' edge. The direction of rotation is, as suggested, leading edge first, and in Sussex is always anti-clockwise when viewed head on. Some millers term this 'right-handed'. The windshaft carried huge timbers called sail stocks or middlings, around 12ins. by 14ins. thick at the centre, and up to 40ft. long; these are also pitch pine. The sweeps are secured to these with bolts and clamp irons. Short wooden braces between sweep, whip and stock called back stays, are added to finish the job. The canvas sail cloth is spread on the driving sides of the sweeps and is furled to present a greater or lesser driving area according to the strength of the wind. This furling is set by rope lines from the leading edge at the tip of each sweep, the cloth being 'taken in' at an angle up the sweep from the leading tip, the basic configurations are known as full sail, first reef, dagger point, and sword point. The biggest drawback with commons is that the miller had to stop the mill and adjust each sweep individually when a change was needed. Also, the initial spreading often required the miller to climb the sweeps to attach the cloth to its heel rail if access from the breast was not provided. Of course, frost tends to aggravate the furling and rain rots the cloth itself, so we come to the shuttered sweep which was a major step forward.

During the 18th and 19th centuries, engineers began to turn their attentions toward the windmill. The rewards for a novel idea or new innovations were great. The names of famous engineers are often mentioned in the context of windmill construction. The name Watt is perhaps best known by the majority. His name is connected with the centrifugal governor applied to the steam engine. This has a thousand and one uses today, from clocks to large marine engines. Stone tentering is the application in the windmill. Thomas Mead was, in fact, the patentee in this context and filed his claim in 1787. Second, is perhaps Smeaton. This man is best known for his work at the Eddystone reef off Plymouth in Devon, where his famous lighthouse stood. He pioneered serious investigation into the power derived by sweeps with respect to size and varying angle of weather. His experiments became standard works on the subject. Then we come to the two that concern us here, Meikle and Cubitt. Meikle, a Scottish engineer, invented what we call 'spring sweeps'. He saw that if the drawbacks of the common sweep could be overcome, the miller would be relieved of a lot of work, giving him more time to attend to his duties inside the mill. This, he achieved in 1772. by having each sweep fitted with shutters in its driving side (Plate 109). These, usually of pine or frames covered with canvas, as at Winchelsea (Plate 116), are pivot-mounted and controlled by a common wooden connecting bar called a sail rod. A large leaf spring is mounted near the inner or heel end of the sweep (Plate 5). An iron rod is connected to this and runs down the sweep to the tip, where it is linked to the sail rod by a setting bar or plate. At this point, the tension of the spring is set according to the wind conditions prevailing. The system here, thus loaded, holds the shutters closed until the driving force of the wind, overcoming the spring tension, spills through the partially open shutters. The miller had to set each sweep individually, bringing them round one at a time with the aid of the brake. We can see that this once set is, unlike 'commons', automatic in its operation. It is not, however, as 'air tight' and a certain amount of power is lost on comparison.

William Cubitt was responsible for the final major refinement to this part of the mill with his mechanism patented in 1807. Cubitt gave his name to what we now call 'patent' or automatic sweeps. His system used Meikle's shuttered sweeps without springs, applying a variable weight counter pressure to the wind through a lever system. This conveyed its effect to all four sweeps at once. We find, when looking at this type of sweep, which can be seen in full working order at Shipley smock-mill (Plates 6 and 7), that at the canister end of the windshaft, there is an iron cross connected through rocking levers or triangles as they are known, directly to the four sail rods. It can be seen that as this spider assembly moves in or out of the windshaft, the sail rods are moved along the whips closing or opening the shutters. From here, the 'patent' gear passes back through the windshaft itself in the form of an iron striking rod. Inside the stone floor of a post-mill or in the cap of a smock- or tower-mill we find that this rod passes out of the tail-end of the windshaft and is coupled to a rack and pinion in most Sussex mills thus equipped. An outside wheel and chain arrangement is connected to the pinion by a shaft (Plates 38 and 99). Here, the miller hangs a weight of perhaps 50lbs. on the operating side of the chain loop (Fig. II). This action, via the pinion and rack, holds the striking rod back towards the tail of the mill, bringing the spider assembly into a position known as 'set for wind' or 'sailed out' with all shutters closed against the wind. The big advantage of Cubitt's gear is that the adjustment for all sweeps is carried out whilst the mill is working. Careful placing of the correct weight on the chain eases the task of keeping the mill speed as constant as possible, provided the work-load remains the same. Two main disadvantages of this arrangement are that, first, unlike spring sweeps, the shutters will remain set open as they come out of the wind when nearing the ground. At this point 'spring sweeps' will close individually, which makes the air-tightness of the patent system less again than that of commons. Second, there being many pivots and bearings in the linkage as a whole, friction causes a lack of sensitivity in its response to the wind.

At Cross in Hand post-mill we find an internally-worked lever system (Plate 9), which, carrying weights, acts directly on the tail end of the striking rod, improving the response. This is described fully under the heading of the mill itself. In principle, the system at Cross in Hand is very like the other main method of loading the striking gear of patent sweeps, namely the 'rocking lever'. When this system is employed, the control weights are hung outside the mill from a series of multiplying levers. This acts on the end of the striking rod with the same effect as the weight wheel and rack assembly described. Stone Cross tower-mill still has this arrangement in place.

The tower-mill at Clayton and the post-mill at Hurstmonceux (page 45) were once equipped with a sweep governor (Figs. IV and IX), so called because it was in the form of a large stone governor which, as the mill speed began to rise or fall due to changing work-load or wind, would operate the striking rod through a clutch and pulley system linked to the weight wheel (fully described on page 46). There is no real record of how successful this particular system was. In principle it seems effective, but required valuable space inside the mill, and very careful adjustment no doubt. The miller who applied this governor, Mr. Hammond of Clayton, did not make his fortune out of it. The windmill was unfortunately at this time in the 1870s, losing ground to the steam-powered mills in any case.

The final form of the sweep in Sussex is the large patent type as seen at Shipley, each sweep 34ft. 6ins. long and 7ft. 2ins. wide, has 36 shutters in 12 bays along its driving side and as many in the leading edge controlled by a very well made set of striking gear. Compare this with the smaller common sweeps at Nutley which, with a pair of spring sweeps to partially automatic, move off in a comparatively light breeze.

Two notes here: at Patcham tower-mill we find that whilst the sweeps are now basic frames, they were 'patents', and instead of having a solid iron spider or striking cross, still carry for us to see a spring-armed component which allowed each sweep a little independence of the other three, making for greater efficiency. This is a variation, fitted by Holloway, of a combination of Meikle's spring sweeps and Cubitt's patent fitted to a number of local mills by Cheale of Lewes. Plate 8 shows Cheale's 'spring-patents' fitted to Southern post-mill at Lewes. The photograph is detail from a plate *c.* 1890 taken by Reeves.

At some Sussex mills, notably Argos Hill and Stone Cross, several of the outer shutters were removable. This was to soften the response of the striking gear on a blustery day. The miller took these out as work load and conditions dictated, thus reducing or increasing the effective size of the sweeps.

The Windshaft

This is the primary driving shaft of the windmill and originally made of wood. Generally between 10ft. and 16ft. long behind the canister, it lies at an angle of approximately 10 degrees to the horizontal in the mill and has two functions. The first is to take power from the sweeps. The second to carry the brake and tail wheels within the mill.

The wooden windshaft of our Sussex mill was originally morticed through for the stocks (last seen at Winchelsea post-mill), but all of the remaining examples carry an iron two-way box casting with a neck and fintail end called a fintail canister or poll end (Plate 188). This is fitted into the windshaft with bolts and hoops and weighs something under one ton.

Iron windshafts (Fig. VI) as a rule have the canister integrally cast on. Barnham tower-mill is an exception. Here the canister is flanged and bolted to the tail, and at West Blatchington smock-mill, we find that the stocks are clamped and bolted to a large iron crucifix in a similar manner to that of the multi-sweep mills of Lincolnshire.

The stocks can be up to 12ins. by 14ins. in section for shuttered sweeps and therefore require a large canister weighing approximately half a ton. To retain the stocks and prevent them from slipping, shoulder nebs or blocks and special wooden packing pieces called poll wedges are employed. Some millwrights fitted pinch bolts as an additional precaution.

Just behind the canister head is a neck bearing. The shaft here rests in a brass bearing cast in the form of a half journal supported by a cast-iron pillow block on the breast beam. At the tail end of the windshaft, we find a much

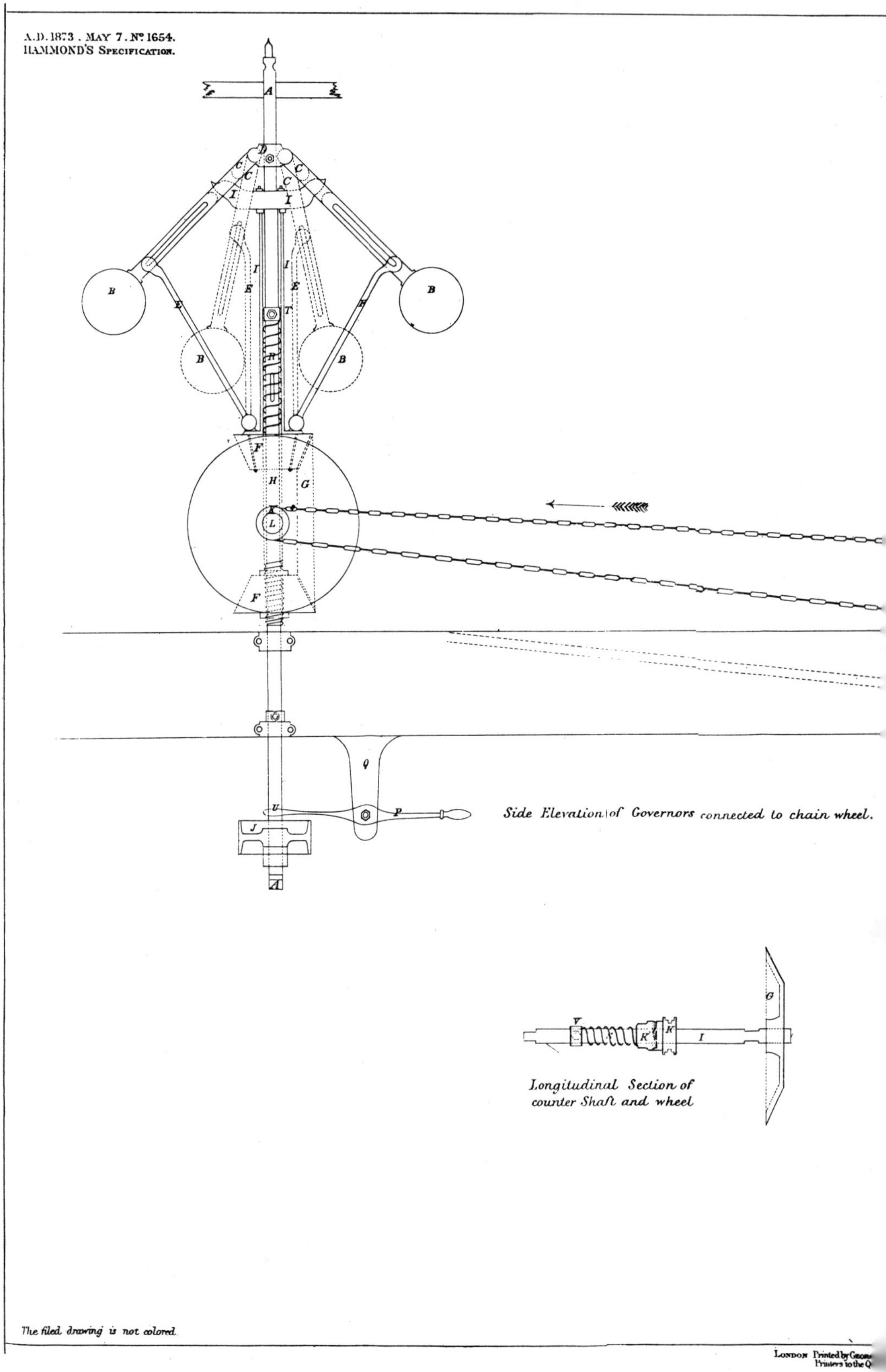

Fig. IV. Hammonds' patent sweep governor as fitted at 'Jack' Tower Mill, Clayton.

(1 SHEET)
Top of Fan Stage
M
N O
Cross Section of chain wheel.
Cross Section of Chain wheel

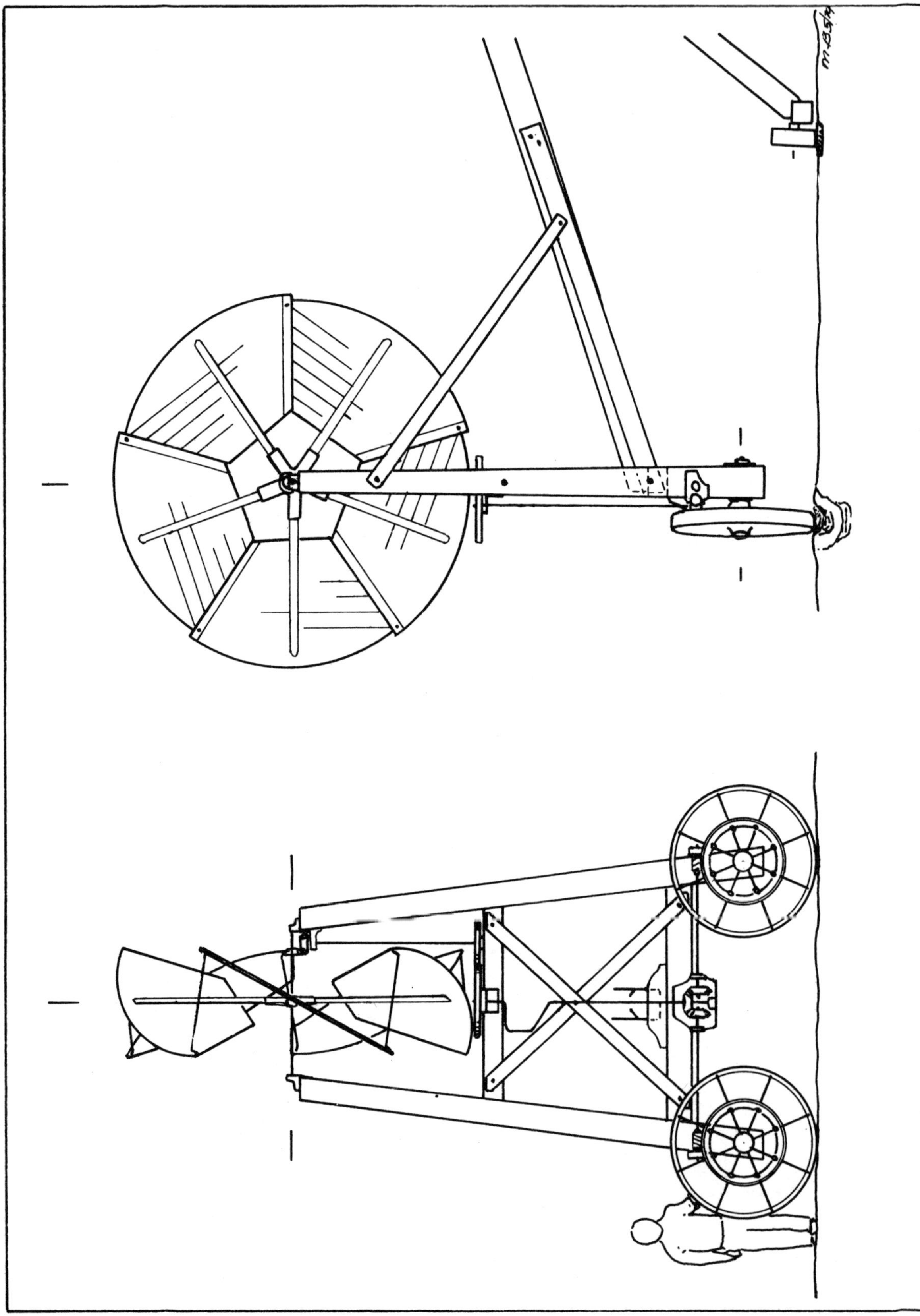

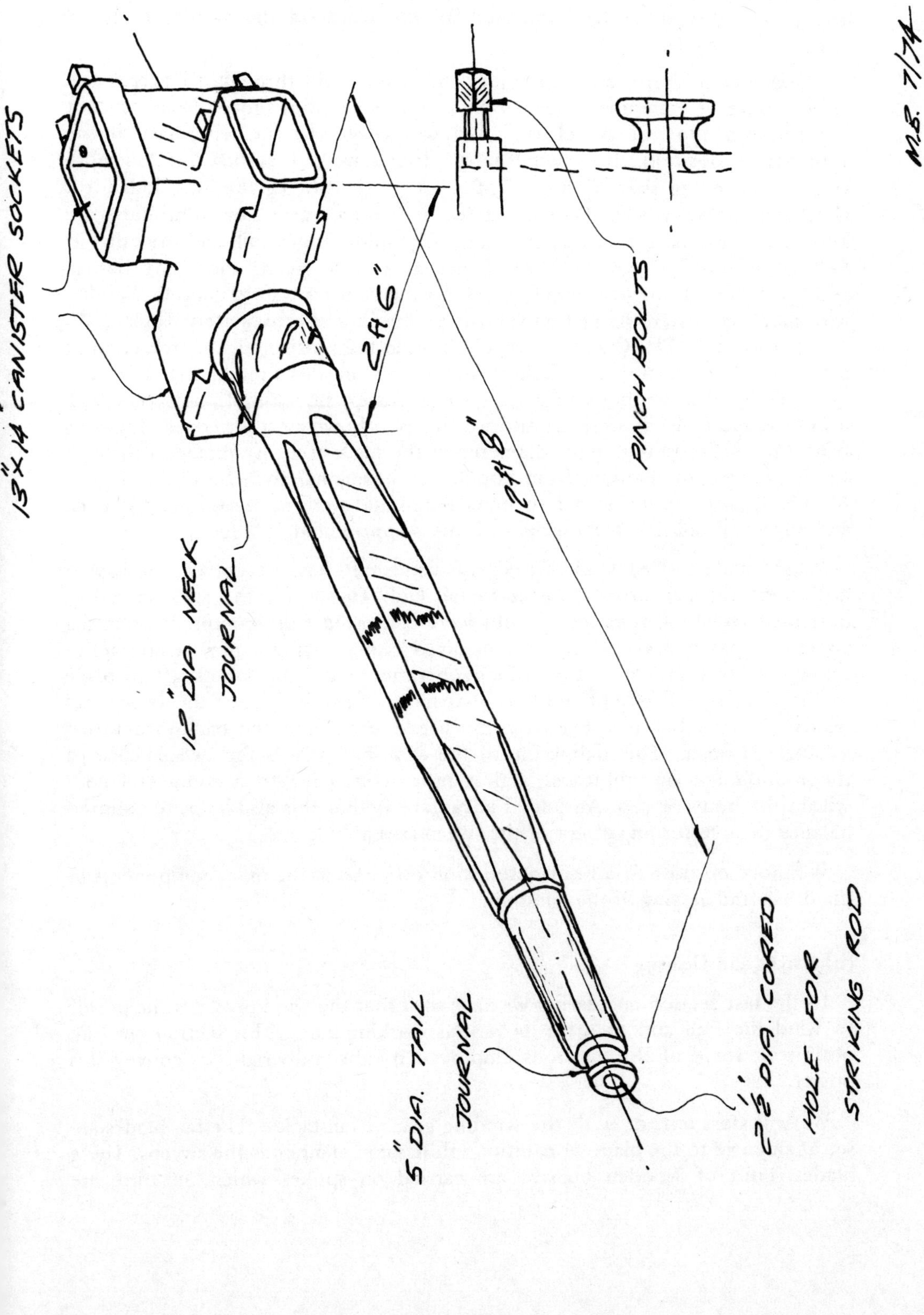

Fig. VI. The iron windshaft from Littleworth Smock Mill.

smaller combined pintle and thrust bearing mounted on the tail beam. This brass takes the thrust load imposed by the wind on the sweeps (reduced somewhat when driving the stones).

Wooden windshafts are sometimes morticed right through to accept the main spokes of a 'compass' arm brake or tail wheel, but more often than not a clasp arm type of wheel is found wedged on to the windshaft at the appropriate position. Iron windshafts often have a cast box-like assembly wedged on to the shaft (Plate 17), the square sides being the ideal hub for a clasp arm brake wheel. Two examples of compass arm iron windshafts are known to have existed in Sussex, one at Clymping smock-mill and the other at Sullington post-mill (Plate 113). Here, as can be seen, the shaft has an additional set of mortice boxes similar to the canister and is unique. Wooden windshafts are often parallel with eight or 12 sides approximately 18ins. A/F, or tapering from 2ft. ∅ at front to 1ft. 3ins. at the pintle end, the iron counterparts are often beautifully fashioned (as seen in the sketch, Fig. IV). The example here is of the windshaft from a Sussex smock-mill. It is very well proportioned and has strength and weight exactly where it is needed. Tapering from the tail bearing of 6ins. diameter to the neck at 12ins. diameter through subtle changes of sections from circular to hexagonal then becoming square for the wooden brake wheel to be clamped and wedged, passing to the neck and huge canister. The total weight of this is approximately 38cwt.

Where patent sweeps are fitted the windshaft has, of course, to have a hole right through from canister to tail end. This is for the striking rod as described above. A number of mills were converted from commons or spring sweeps to patent sweeps and this involved either fitting a new windshaft or a very tedious and accurate drilling operation with the windshaft in place in the mill. It will be noticed here that most of the weight of the sweep and windshaft assembly is carried by the breast beam and the backward thrust by the tail beam. The inclination of the windshaft places the sweeps clear of the roundhouse or mill-tower and is more conducive to a compact layout within the body or cap. An added advantage is that this also helps to counterbalance the overturning effect of the sweep assembly.

We move on now to a brief description of some of the main components in the drives and gearing of the windmill.

(ii) Drives and Gearing

In the last section and earlier we have seen that the wind provides the power to wind the mill and to drive its various working parts. This section gives an outline of some of the methods employed by the millwright to convey this power.

We will start outside with the winding gear or fan-tackle. The fan blades are set at an angle to the plane or rotation in the same manner as the sweeps. These blades, built of wooden boards, are carried on spokes which, in turn, are secured to a fanstar and shaft. The rotation imparted by the wind, is transmitted by bevel gears and through a spur gear reduction to the drive shaft itself. When fitted to a post-mill, the final drive comprises two large iron winding wheels 4ft. to 5ft. in diameter. Each of these has a gear ring engaging with worm shafts. These are driven by bevel gears meshing with the lower end of the drive shaft. Two examples of this can be seen on the fanstages at Cross in Hand (Plate 11 and Fig. V), where over 1,500 revolutions of the 10ft. 9ins. diameter fan are required to turn the mill through 180 degrees, and Argos Hill post-mill (Plate 12), where large eight-bladed fans are used. At Icklesham we find that the smaller fan is on top of the mill body (Plate 13). Here, the principle is the same, but the drive shaft is in three pieces interconnected by bevel gears. The top two lengths run down the mill body from the fan and mesh with the third, engaging with the worm shafts at the bottom of the steps (Plate 14).

For smock- and tower-mills, the arrangement is very similar in that the worm shaft becomes a large worm pinion proper (Plate 16) and instead of engaging with winding wheels it meshes directly with the rack which is formed as part of the curb on the tower itself. The rack is, of course, a fixed part so that the winding gear pulls the cap around it and into the wind (Plates 15 and 16). On older smock and tower-mills, wooden worm and racks are used, as seen at West Chiltington smock-mill and East Wittering tower-mill (Plate 159), iron components becoming the rule later.

As mentioned earlier, the original method of winding in the case of the post-mill was to pull the mill round its post using the tailpole (Plate 10). Smock- and tower-mills also had a tailpole which, braced to the cap, hung vertically within the miller's reach. This served the same purpose. An intermediate method, still to be seen on Halnaker tower-mill, is the hand winding wheel and chain. In this case, the winding wheel is greared directly to the rack. In others, the worm pinion was driven by a similar wheel and chain.

Inside the mill, the windshaft carries the brake and tail wheels. When made of wood these are often built up of oak or pitch pine in compass or clasp arm configuration. The former being a rim of wooden felloes supported by spokes which pass through the windshaft, morticed to accept them. The latter, favoured by later millwrights, is built up of a similar wooden rim and a box-like centering. The middle aperture of this is wedged over the windshafts or fitted on to a complimentary square locating hub. Iron brake wheels are common, usually cast in two pieces and clamped together over the windshaft. The cogs fitted in the inner face here are of apple wood or hornbeam (Plates 17 and 18). Iron wheel rims have cast mortices for these. Wooden wheels have hand-cut mortices. In all cases, each cog is hand-fashioned and has a stout shank. This is fitted and driven into its mortice, the shank held with a steel pin. The gearing in a windmill is generally surprisingly quiet when running. This is due to the skill of the millwright in matching each cog with the mating wheel or pinion. Iron meshing with iron was not favoured because of the noise produced and also the tendency of this combination to wear badly. We usually find that brake and tail wheels have wooden cogs and that the drives to the stones which engage with these are of iron.

The stones are driven in a number of ways as described earlier under mill types. The overdrift arrangement in a post-mill is the simplest. Here, the runner stone is driven by an iron stone nut and quant spindle (Plate 23) which meshes directly with the brake or tail wheel. The cogs of the stone nut here are slightly bevelled because of the angular displacement of the windshaft and the method of driving from the inner face of the wheel. In spur-gear post-mills, as mentioned earlier, the stones are underdriven via a wallower or crown wheel and a short upright shaft, the stone nuts and spindles being driven from a spur-wheel which has cogs around its circumference (Plates 25 and 26). Smock and tower-mills have a similar arrangement (Plates 21 and 28). The great spur-wheel is driven by a long upright shaft topped with an iron wallower (Plate 27) meshing with the brake wheel above. The shaft here is of oak or iron. Other machinery such as dressers, cleaning machines and oat rollers, are lay-shaft skew-gear driven off the brake and tail wheels or by belt and pulley. In smock and tower-mills this auxiliary drive is via a spur drive meshing with the great spur-wheel or, as at Barnham, a separate lower bevel wheel (Plate 22). In latter days, provision was often made to take power from an engine in still weather. A number of slip cogs were removed from the brake wheel and the mill driven from below. At Shipley the windshaft can be jacked forward 'out of house' to allow this. In a post-mill, the sack hoist may be friction driven by belt or chain from the windshaft, or from an intermediate shaft gear driven from the back of the brakewheel and via a friction clutch or bevel wheel below the wallower in smock and tower-mills (Plate 35).

I mention that the brake is usually operated by lever and rope. However, Holloway fitted a screw-down brake, applied by wheel and chain to several mills (page 83).

Of the stones themselves, it should be said perhaps, that whether over or underdrift, the top or runner stone is always the driven stone, rotating over the lower or bed stone. The drive has, of necessity, to be imparted as a rotation without communicating any bias or strain to the finely-balanced runner stone. This is the most accurately set part of the mill's motion and has to be, especially when producing first-grade wheat meal for flour. The runner stone is driven and held off the bed stone by a spindle which passes through a bearing in the bed stone. This bearing may be brass, but is often of a wood such as holly or lignum vitae. The vertical setting of this stone spindle is controlled by the tentering gear, described in the next section.

The drive is communicated in one of two ways. First, by bridge and mace. In this case an iron bridge or stone bar is set across the eye of the runner stone. This is turned by the drive dogs of a mace iron on the stone spindle. The entire weight of the stone is carried by a conical nib—or 'tomkin' in Sussex—at the centre of the mace. The second method is to carry the runner stone weight on a gimbal iron, two opposing gimbals supported by the stone spindle, the other two in the stone itself. This method (not as yet known to the author in Sussex

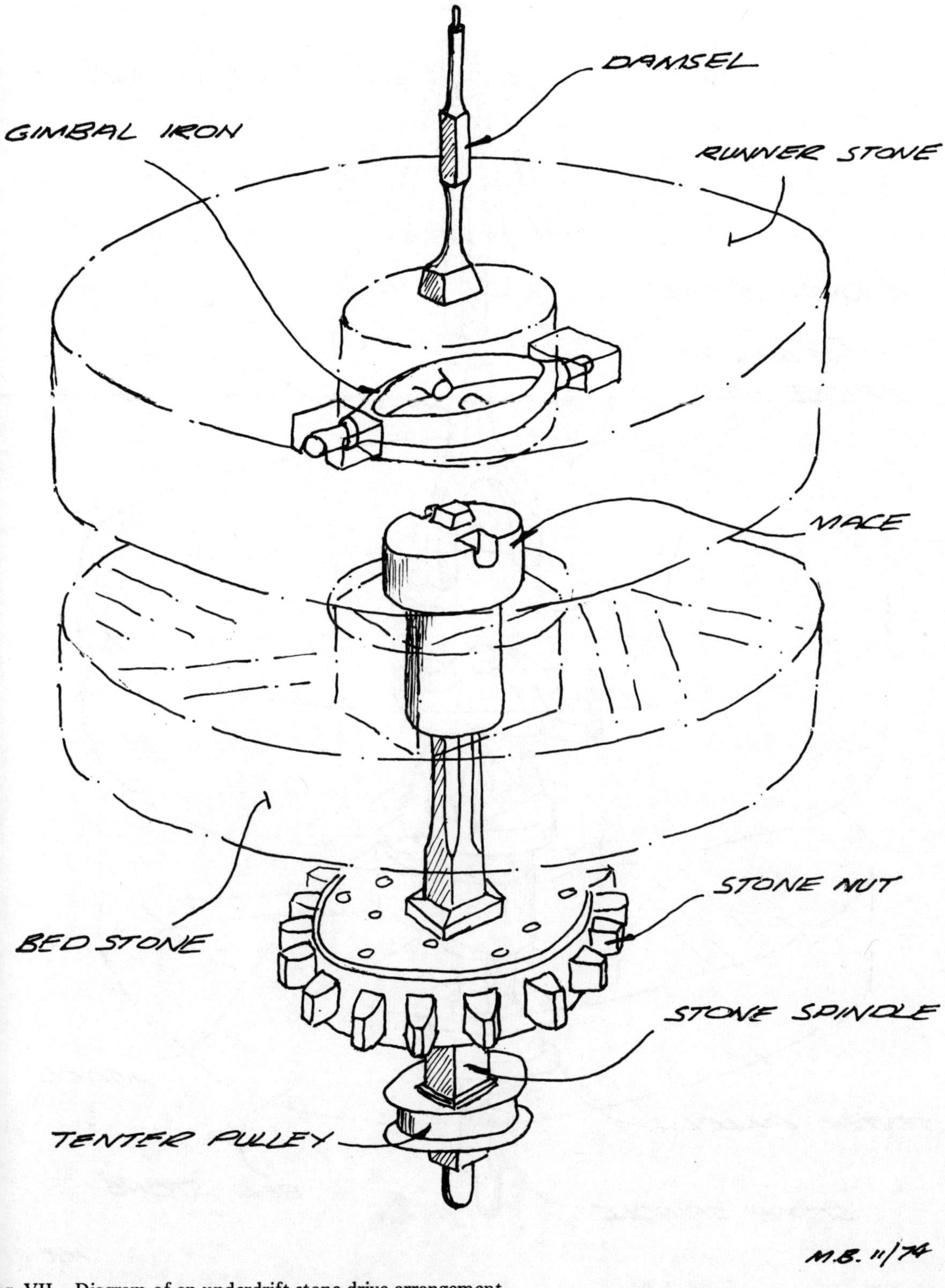

g. VII. Diagram of an underdrift stone drive arrangement.

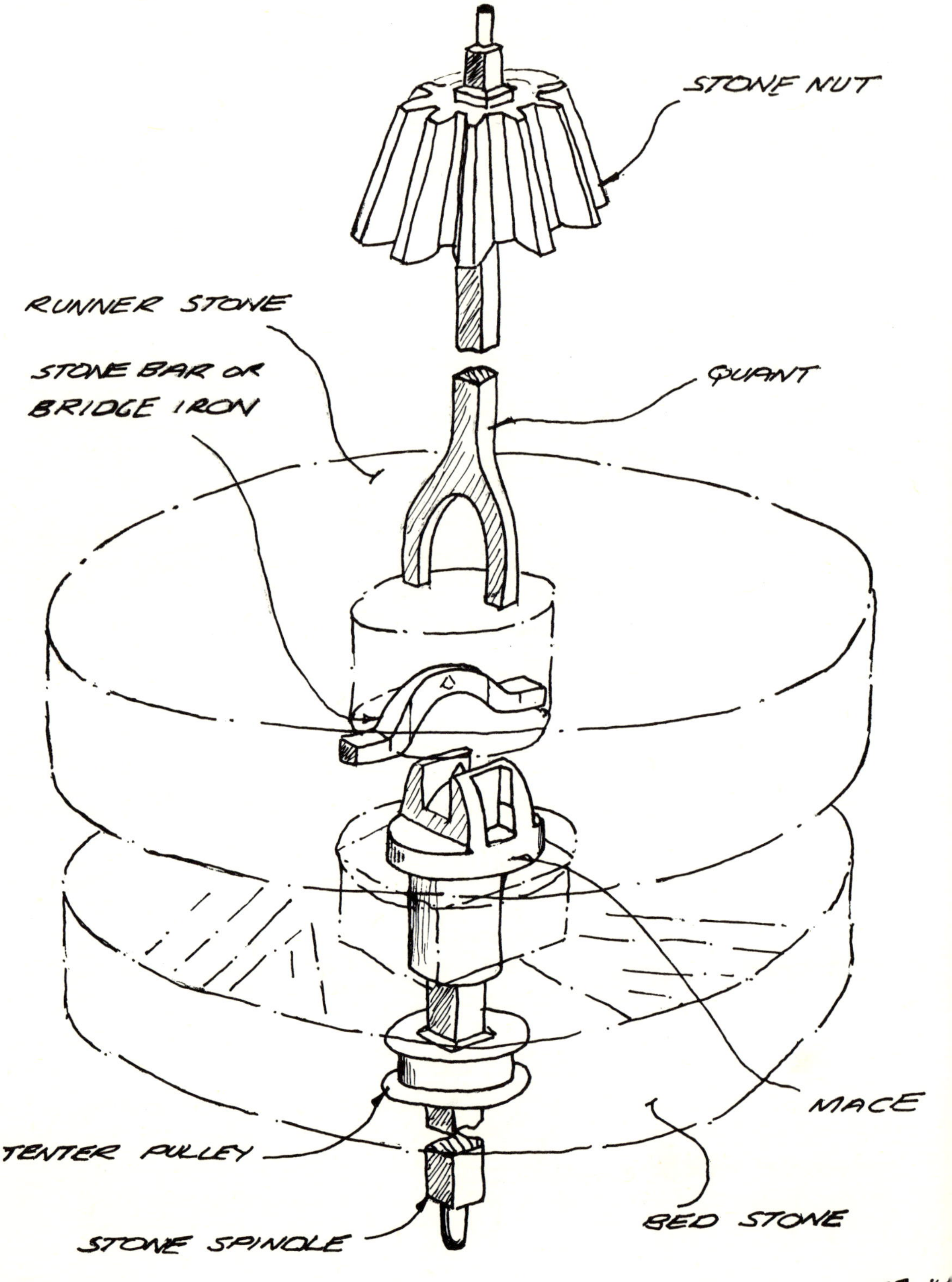

Fig. VIII. Diagram of an overdrift stone drive arrangement.

mills), whilst being more independant of the drive, tends, I believe, to impart a lift to the runner stone with an increase in load. To achieve a full understanding of the combined support and drive of the runner stone from the written word is difficult, so I have included sketches of the two methods described (Figs. VII and VIII).

(iii) Stones and Tentering

We come now to the stones themselves (Plate 30). The action of the runner stone rotating over the bed stone, whilst grain between is reduced to meal is known as grinding. This is the combined result of crushing by the sheer weight of the runner stone and shearing by a scissor-like action imparted by a traditional pattern of grooves and furrows in the surface itself. In order to achieve this result, the stones have to be heavy. A new 4ft.-diameter peak stone weight approximately 18cwt., and the same size French burr about 17cwt.

The faces of the two stones are sharpened or dressed with radial and tangential furrows and grooves. Between these is a pattern of cracks or stitches (Plates 30 and 170). The direction of the radial pattern is handed so that the grain is eased outwards across the face of the stone. The stone dresser uses hardened steel cutting tools to sharpen and dress the stones (Plate 171), namely the mill bill and pick. The bill, weighing 3½lbs. when new, is flat-ended and used for levelling the surface generally, cutting the furrows, and cracking or feathering the lands; the pick, usually made from a bill which has become too light, for stitching or pitting. These tools are wedged into a wooden holder called the thrift. A very informative first-hand account of stone dressing by the late John Russell, miller at Cranbrook smock-mill, Kent, is to be found in *The English Windmill* by Rex Wailes.

When the runner and bed stones are face up, the dressing appears identical so that when the runner stone is face down and rotating over the bed stone, the patterns are opposed and the action similar to that of shears or scissors. Stones are dressed to run either 'with the sun' (clockwise) or 'against the sun' (anti-clockwise). This is, of course, pre-determined by the hand of the drive within the mill, e.g., overdrift stones (Fig. VIII) in an anti-clockwise post-mill run against the sun.

The dressing varies according to the type of work being done. The grinding of wheat for flour is very demanding and requires well-sharpened stones. In this case, the bed stone is often flat and the runner stone face, or breast, has a very shallow dish to it to encourage a progressively finer action from the eye outward towards the edge or skirt. As mentioned earlier, the two stones should never touch, the gap at the skirt for flour-grinding being literally a hair's breadth when running properly 'tentered'.

Two main types of stone are found in Sussex mills. First, peak stones (Plates 30 and 170), which are of millstone grit, a type of hard sandstone quarried in the north of England (Peak District). These stones were mainly used for grinding barley, oats and maize. Second, French burrs (Plate 30), this type being easily recognisable in that it is built up in interlocking segments which are set in a supporting plaster of Paris base. The segments are of a type of fresh-water quartz and came mainly from France, thus giving this type of stone its name. (Stones of this type made of English chert have been found, but are rare.) In nature, the quartz was shaped in smallish pieces and has a definite grain. The stonemason, therefore, had to match and fashion these pieces to make a suitable matrix.

The big advantage of the French burr (used primarily for grinding wheat for flour) is that it is very sharp and can be given a fine dress which will not contaminate the flour with sandy particles (page 162). In addition to these traditional stones, the cullin stone is sometimes found. This is a continental equivalent of the peak stone and blue-grey in colour. More commonly though, the composition or emery-faced variety which came into use in the later stages of stone milling.

All French stones are bound with iron hoops for safety, the runner stones are balanced both statically and dynamically for perfect running by adding lead or weights at the appropriate points on the back or circumference. Bed stones are levelled on their beams by careful wedging underneath to match the pivotal plane of the runner stones. The stone spindle has to be set exactly vertical to ensure that its axis is perpendicular to the level bed stone. This is attained by securing the 'Jack Stick' to its head. Plate 172 shows the one used at Cross in Hand. This is clamped to the stone spindle by means of a locking wedge and has a feather quill in the split end. The spindle is rotated and its vertical attitude adjusted at the footstep bearing so that quill grazes the lands of the stone equally around the circle described. This is a patently simply craftsman's micrometer giving exactly the result required.

Above the stones and mounted in the horse frame, we find the hopper (Plates 23 and 29) which is kept charged by chutes from the bins. Below this is the shoe. An agitator, called the damsel, shakes the shoe and thus the grain into the eye of the stone at a rate regulated by the miller's feed line. This terminates at the twist peg, near the meal bin (Plate 28) in the spout floor below, to adjust the height of the shoe. A small bell alarm is often set up here so that when the hopper is nearly empty it is made to ring, thus warning the miller (Plates 23 and 29). This saves the stones becoming damaged or firing the mill.

The stones themselves are enclosed either in wooden casings, which are octagonal and made of boarding, or round wooden 'tuns'. These are barrel-like in construction and hooped with iron bands. Ground meal emerging from the skirt falls into the annular gap inside the casing and is then swept round to the spout aperture. When driven from below or 'underdrift' (Fig. VII), the stone spindle carries the stone nut and takes the drive through the bed-stone bearing to the runner stone above.

More often the arrangement in the windmill is overdrift (Fig. VIII). Here, the stone spindle is driven from above by a crutch-pole or quant which straddles the drive to the runner stone without interfering with it. This spindle is always the final drive to the runner stone, whether over or under drift.

Beneath the stones in the spout floor is the tentering gear (Plates 28 and 31). Here it can be seen that the stone spindle is supported in an adjustable brass footstep bearing. This bearing takes the total weight of the runner stone. The tentering gear controls the vertical setting of the spindle via the bridge iron. The bridge iron has a fulcrum at its anchor point, the stone spindle bearing in the middle and at its free end, is itself carried by the brayer iron. The brayer has a fulcrum at its anchor point, the bridge iron bearing on it and at the free end, the tentering screw for hand setting. The mechanical advantage at this point is about 6 : 1 in this example. To one side and driven by a belt, is the stone governor, used to automate the setting. This is of the centrifugal type and when running at the correct speed the rotating weights raise a steel yard some 6ins. from rest. As is seen, the steel yard supports the tentering screw with a leverage of approximately 30 : 1. Therefore the total advantage through this lever system is approximately 180 : 1. At rest, the system raises or 'lightens' the runner stone off about one-thirtieth of an inch, the weights being quite sufficient to do this through the governor linkage itself, thus allowing an easy start when the brake is lifted. As the mill speed rises to its correct level centrifugal force at the weights causes them to move outwards and raise the steel yard until the limit of the linkage or the stop collar is reached. This action lowers the brayer and bridge irons and, of course, the stone spindle, bringing the runner stone to the correct nip, this being previously set by the miller at the screw or handwheel.

(iv) Cleaning, Grading and Dressing

Cleaning machines do just that, i.e., remove loose husks, chaff, dust and mouse dirt, usually by blowing air through the grain as it passes through a sieve or centrifugal grader. Smut and other fungus is removed to some extent by similar means. Here, the grain is rolled vigorously in a vertical air-blown rotating sieve. This machine is called, naturally enough, a smutter (Plate 33). The cleaned grain is often passed on to a separator which actually separates the large grain from the smaller, and then put through the grinding process. The resulting meal gravitates to the bins on the spout floor.

In the ground form (referring to wheat, of course), we have wheatmeal which is the whole ground wheat grain, including the starch kernel, the outer bran layer, and the germ. This meal can be used for wheatmeal bread and biscuits. For finer grading, the meal, usually after being allowed to cool, is passed back up to the dresser (Plate 34). This is also known as the flour or wire machine and takes the form of a rotating cylindrical wire mesh sieve through which the meal is passed. The mesh of this drum increases in size towards its open end. As the meal passes through this it is brushed against the walls and falls into separate spouts for each grade. There are three main grades. High-grade fine flour or 'firsts,' coarse flour or 'seconds'. The remaining bran and larger meal particles are called 'thirds'. The thirds are separated into pollards and bran. The first and second grades are used for baking purposes, the rest usually for other food products and animal feeds. In earlier times when there were many feudal mills still operating, tenants would bring their grain to the miller, who took a portion of the meal as payment for work done, in turn, some of this would be taken by the landowner as tithe payment. In later days the miller kept the pollards back for his own livestock, or perhaps to sell elsewhere.

In the more up-to-date mills, all the above machinery was present. In some cases today we find that the dresser survives along with an oat crusher and kibblers for provender working. The latter were used for cracking wheat or maize grains into three or four pieces for feedstuffs or maize prior to grinding. Peak stones were sometimes used to crack field beans in a similar way. The smaller mill would have had a dresser only or its predecessor the bolter. This had a cylindrical cloth sieve and performed a similar, if less demanding, function to that of the dresser. All these machines were driven by auxiliary drives. In the post-mill, the power was usually taken from the brake wheel. The smock- and tower-mill layout provided much more room and there was sometimes a separate floor which contained all the cleaning and dressing machinery called the upper spout or dresser floor.

(v) Flour

Here is an aside for the reader to consider. Present-day white bread is often made from foreign hard wheat. This is mainly Canadian, which crushes very well and will absorb a higher proportion of water than English. In huge dockside mills, the wheat is today break-rolled into a flour, not ground. The resulting meal is systematically split up into its constituants which are all used for different purposes. Bread flour today is the starchy kernel minus most of the natural goodness and roughage of wholemeal flour. The wheat germ and its accompanying vitamin layer are cracked out separately and like traces of bran removed.

The starchy remnants are literally bleached because the consumer is said to prefer whiter bread. A calcium filling agent may also be used (akin to chalk, I believe), and chemically-reproduced vitamins added to the mix, legislation making this compulsory.

Brown bread is often made with a compromise white flour with some of the natural subtractions added in again, plus a colouring agent. The main types being wheat germ and wheatmeal. Wholemeal is the only type today that resembles the original home-baked bread. This was made from stone-ground flour before the steam-powered roller mills began to overwhelm the windmill during the late 19th century. The paradox here is that wholemeal bread, which is made from flour that has not had to go through any of the purging, enfeebling and additive processes than white bread flour has endured, is more expensive than its counterpart!

(iv) Working Speed and Power

The power developed by a windmill is somewhat difficult to determine precisely, owing to the variable nature of its working. However, I include several examples of calculated horse power to give some idea of the work done by these natural power engines. The figures are approximate.

The wind is the first factor considered here. The pressure exerted by moving air is proportional to the square of its velocity. For example, the driving force of the sweeps when working in 20 m.p.h. wind is four times that from a wind of 10 m.p.h. A constant factor is friction. The wind has, of course, to overcome the lethargy to move due to friction before any useful work can be done. Millstone speed itself varied from mill to mill and for different types of work, between 100 and 150 r.p.m. We find that the later conventional mills are geared at between 11:1 and 7½:1. We know that 4ft. French burrs were run at approximately 125 r.p.m. We can, therefore, see that a working sweep speed of 11½ to 16½ r.p.m. corresponds. Using vectored force and these speeds in a basic power calculation, the following figures may be obtained:

A small post-mill running at 15 r.p.m. sweep speed would develop up to four-sweep horse power in a 15 m.p.h. wind, and approximately 12 in a 25 m.p.h. wind using common sweeps. A large post-mill like, for example, Cross in Hand running at a similar speed in a 15 m.p.h. wind may well have mustered up to eight-sweep horse power, and using her patents in a strong wind of 30 m.p.h., approximately 32 horse power.

The inefficiency of the working mill would, of course, reduce the power available at the stones somewhat. Multi-sweep mills were improved in the 19th century to take full advantage of the wind, and Sussex had its own six-sweep post-mill. This stood at Ashcombe, near Lewes, and was an imposing sight (Plate 104).

The power developed by a multi-sweep (i.e., more than four sweeps) is not directly proportional to the number of sweeps. This is due to the increased turbulence created in the air. However, it seems likely that Kingston mill with six large double-shuttered patent sweeps was potentially the most powerful post-mill ever built in Sussex at around 45-sweep horse power in a strong wind. It is interesting to note the design of her sweeps. The millwright had employed a good angle of weather at the heel and equipped the inner bays with larger shutters than those fitted at the tip, perhaps in an attempt to improve the wind flow. The main advantage of the multi-sweep mill was not so much that it had higher potential power, rather that it was able to make better use of light winds. When working in a light wind, the miller would start using dressing machinery or one pair of stones at the correct speed. The other stones and machinery would be brought in if and when the correct condition prevailed.

Generally a four-sweep mill would start to run in a light wind of 10 m.p.h. The best working wind was considered to be around 20 m.p.h., but mills with shuttered patent sweeps could be controlled in a wind up to 40 m.p.h., so it is said. Above this, the brake would be applied and the mill stopped for safety's sake.

I note that the post-mill at Bexhill Downs had a pair of 4ft. 6ins. $\emptyset$ French burrs which were geared at 6 : 1, and a pair of 4ft. 3ins. $\emptyset$ peak stones geared at 7½:1. The stone speed here would have been approximately 95 r.p.m. for the burrs and 120 r.p.m. for the peaks. This reflects the earlier practise in windmilling of running a large stone at a slower speed to obtain the required result. As mills progressed, stone speeds tended to increase as the size lessened.

Finally here, the following notes of power required may help to put the foregoing power ratings into perspective and are offered as locally accepted by experienced men. The state of dress of the stones and condition of the grain made a considerable difference to the working, as, of course, did the wellbeing of the mill itself:

i. Dressing and light grist work: up to 4 horse power.

ii. One pair of stones grinding one sack of wheat per hour for fine meal: 6/8 horse power.

iii. One pair of stones grinding one sack of barley per hour for fine meal: 10/14 horse power.

iv. One pair of stones grinding one sack of oats per hour for Sussex ground oats: 12/16 horse power.

One sack = 4 bushels, i.e., 252lbs. wheat, 224lbs. barley, and 168lbs. oats. At Cross in Hand, it was considered good going to grind at the rate of two sacks of wheat or one of oats per hour with a good wind.

When going over to engine power, windmillers found by experience that plant rated at 10 or 12 h.p. could be employed to drive one pair of stones, but 20 h.p. came to be regarded as the most realistic alternative. Power mills equipped with engines rated at between 40 and 50 h.p. were expected to work with two pairs of stones, plus cleaning, dressing and mixing machinery.

CHAPTER FOUR

THE SETTING

BEFORE MOVING on to the mills themselves, a few words here about the county that nurtures them—one of the bigger counties of England, Sussex has as much, if not more to offer than most. The English Channel forms the southern boundary. Eighty miles of shoreline varies from high chalk cliffs to shallow shingle beaches with a sandy bottom. The latter are in the majority, where the coastal plains and marshes meet the sea, here also are several fine stretches of sand. We find a number of windmills remaining near the sea along the coast where they had to stand the full brunt of Channel storms.

Moving inland, we encounter one of the finest ranges of chalk hills in England, the South or 'Sussex' Downs. Their southern slopes rise at a gentle but deceptive rate from the sea. These reach a neight of over 800ft. at several points and stand bareheaded for the most part, unlike the wooded North Downs. From these heights, where the first inhabitants of Sussex lived, the northern scarp drops steeply into the Weald. The geology of these hills and the rest of the county shows us that at one time the gentle seaward slope continued upwards to a 2,000ft. peak[1] over the High Weald, falling again to join the North Downs, the whole forming a Super Down. Aeons of weather, including ice age thaws, have laid bare the sandy lower strata of the Weald. This rises to 800ft. in Ashdown Forest. As with the coastal plain, a number of mills still survive on the Downs, Jack and Jill at Clayton being outstanding examples. These fine chalk hills offer wonderful views over the sea and the Sussex Weald, from exhilarating walks on fragrant turf. Perhaps the most impressive is that from the high point at Firle Beacon. On a clear day in summer, this has to be seen to be believed. The English Channel forms a steel-blue southern horizon stretching from Rye Bay to the Isle of Wight, and inland Kent is seen to the east, linked by the Wealden Ridge to the hills of north-west Sussex. Those with an eye for windmills, will seek out and find those at Chailey, Mayfield, Cross in Hand, Punnetts Town, Battle, and Stone Cross without difficulty. The recently much-publicised South Downs Way runs the length of these hills. Stone Age man knew this well, and it is today very popular, its earnest walkers are much in evidence.

The Weald itself forms the heart of Sussex and has the most surviving windmills. It is still a mainly agricultural area, though being slowly corrupted by metropolitan overspills and light industrial expansion. Here, the South Saxons made their homes in the thickly wooded parts of the High Weald. Its deep coombes carry the headwaters of the county's rivers. Many watermills made use of these for iron workings, the wool trade, paper-making, and, of course, flour and feeds. The fields and pastures that we see today in the Low Weald were once as thickly forested as parts of the Wealden Ridge are today. This was the Saxon 'Andredesweald', from the Roman 'Sylva Anderida'.

The whole county is full of natural beauty and splendour and a witness to English history from the Stone Age through Roman, Saxon, and Norman invasions, and settlements to present day. The ancient county towns and villages full of history and character, saw the birth of the iron industry with its mill ponds and water-driven hammers, and furnaces long since cold. The heathy expanse of Ashdown Forest today bears witness to the country's dependance on English oak for shipbuilding. Little wonder then that some of the finest windmills in England stood and stand here today. As mentioned there are windmills and remains to be found in the three main geographic areas in similar proportions. These will reward the lover of the English country scene and the student of industrial archaeology. I suggest that the reader should, when visiting the windmills of the county, spend time taking in the various local aspects that are to be found about each site, for no two are the same in the County of Sussex.

CHAPTER FIVE

THE MILLS TODAY (Fig. XI)

HAVING DEALT with basic construction of the county's mills and traced a brief outline of the internal economy, we come to the heart of the matter, i.e., the mills themselves. The windmills of Sussex are engagingly varied. No definite pattern for any of the three main types is found, although many have similarities. Post-, smock- and tower-mills are well represented in similar numbers. The actual age of the mill itself determines a lot of the main differences that we see today, though millwrights over the years changed their methods and ideas somewhat. The purpose of the mill often played a part in the form it took. We find that the business mill became quite specialised in later times and that the earler form, often feudal, was as simple and effective as possible. The latter was the first to decline as roads and transport made it wasy to take grain to a modern well-equipped mill. However, examples of both still remain. Cross in Hand post-mill, Shipley smock-mill, and Polegate tower-mill today representing the former, Nutley post-mill, West Blatchington smock-mill, and Halnaker tower-mill, the latter.

Around 1820 Sussex had perhaps half-a- dozen windmills for every one standing today. The oldest standing now is perhaps High Salvington post-mill which dates back to the early 18th century. Nutley post-mill has the appearance of a genuine early 18th-century mill, although this is only conjecture. Icklesham post-mill, which has often been rather extravagantly claimed as late-17th-century or early-18th-century, cannot be positively traced before a notice of sale in 1781. Of the smock-mills, Rottingdean can be positively dated 1802, and Somerley mill at Earnley is very close to this. Halnaker holds the honour for the tower-mills at *c.* 1750. At the other end of the scale, Patcham tower mill is definitely the youngest in Sussex, and one of the last to be built from scratch in England for flour or grist working. This started life in 1885. Shipley smock-mill is the runner-up, so to speak, built in 1879.

Sussex windmills today still have features which are attributed to the county. The reader will find these mentioned here and earlier, the most notable being perhaps the large stage-mounted fans fitted to the tailpoles of later post-mills.

This chapter is devoted to mills standing today. Those familiar with the Sussexian's landscape will know that its mills vary considerably in their internal and external economy. This makes the task of distinguishing between mills proper and 'remains' a little difficult and arguable at times. I have, therefore, classified any recognisable mill structure that still carries a windshaft or stocks as worthy of a place here. Such as windmill structures without these, post-mill roundhouses, smock-mill bases, trestles, windshafts and bygone mills of particular interest, merit a place in Chapters Six and Nine. The lists on pages 182 and 183 will clarify this.

The mills are taken in alphabetical order of type and at the end of each note, the Ordnance Survey Grid Reference is given to help location. Most of those in this chapter are shown in the One-Inch Series of the Ordnance Survey map. The revised 1:50,000 Series, however, shows them all as a standard symbol with or without sweeps.

I

POST-MILLS PROPER

'JILL'—CLAYTON (Formerly Lashmar's New Mill)

White Post-mill having had Sussex Tailpole Fan-tackle

This white post-mill is elder sister to the black tower-mill close by appropriately called 'Jack' (Plates 36 and 38). These two stand on Duncton Down, rising above picturesque Clayton village, and have greatly contributed to the popularity of this local downland beauty spot.

Not far off, the South Downs Way passes by and the walk from here to Ditchling Beacon, two miles east, affords superb views over the mid-Sussex Weald to the North Downs, and also to the sea beyond Brighton, once the home of Jill.

This mill was brought to Clayton from a site above the present-day Belmont railway tunnel and close to Dyke Road, where she was known as Lashmar's New Mill. A painting by Nash and dated 1839 in the Royal Pavilion Art Gallery, Brighton, shows her standing above the newly-constructed eastern portal of Belmont tunnel. Similarly, an engraving in *Handbook to Brighton 1847,* shows her as the most southerly of the three Dyke Road post-mills. Both of these illustrations clearly indicate that the mill originally had a top fan arrangement, as did Cross in Hand before her removal, and Icklesham today.

The date of Jill's erection is understood to be 1821, when she replaced Lashmar's 'Old' mill. She was brought over the Downs to Clayton on a trolley sledge, dragged at first by horses and then oxen.

The route is very difficult to establish, but no doubt the whole was taken down into Preston, from where the journey would have been up the gently sloping valley to the top of Clayton Hill. The date has now been established, not without difficulty, as various maps give conflicting evidence. William Figg's map of 1861 shows Duncton mill at Clayton and Lashmar's mill still in Dyke Road, but more reliable plans of Brighton leave little doubt, placing the move between 1850 and 1853.[2] The first map to show her at Clayton was surveyed in 1873[3] and shows Jill, Jack and the roundhouse of the old Duncton post-mill (see page 96). Mr. James Mitchell was the miller at the time of Jill's removal and worked both post-mills before Jack was built to replace the older one *c.* 1866. (An extract from Mr. Mitchell's 1849 ledger is to be found on page 97.)

A similar mill removal in 1797 is commemorated by a contemporary etching in Brighton museum. This shows a post-mill being drawn by 86 oxen supplied by local farmers for the purpose. This move was from a site near Regency Square to a spot two miles north, above Preston village. It has been suggested that this became the earlier Lashmar's 'Old' mill, but there were several mills on Dyke Road at this time, and the Belmont site is only one and a half miles away from Regency Square, which makes this rather doubtful. There were, in any case, a great many mill removals in Brighton as new buildings took away the miller's wind. One trolley/sledge was reputed to have been used three times each for a different mill.[4]

Hedges now surround the site, and between Jill and Jack is a modern private residence. This bungalow stands on the site of the granary and mill buildings used by the late-19th-century millers, Messrs. J. & C. Hammond, who were here between 1871-2[5] and 1903.[6] J. Hammond worked the giant post-mill at Hurstmonceux during part of the same period. The roundhouse of Duncton mill still remains and is part of the owner's home close to the base of Jack's tower.

Whilst working, the mill was driven by patent sweeps carried by a wooden wind shaft with an iron canister, which bears the inscription I.L. 1831, perhaps commemorating the change to patents. These were controlled by an outside weight wheel and chain. The present sweeps are basic frames and were fitted by Paris of Hove in 1971.

Two pairs of stones were overdriven via head and tail wheels. The clasp arm tail wheel here appears to have been used as a compass arm brake wheel at some time, and it still bears the witness left by a brake around its rim. Perhaps this wheel served in the mill as such in its earlier days in Brighton. The head stones were peaks and are now gone, removed to lessen the load on the floor. A pair of French burrs are still in place under the tail wheel. Of the other machinery, portions of the tentering gear, the flour dresser, and sack hoist remain on spout, stone and bin floors respectively. The bin floor itself is large and the two main bins apparently served the stone hoppers via long wooden feed chutes.

An interesting point to be noticed in the spout floor is that the post which is crowned with an iron pintle is built up of four pieces of timber[9] held together with trenails and iron bands forming the whole, almost 2ft. 6ins. square at its horn ends nearest the cross trees. Below, the part brick roundhouse is deeply sunk into the chalk to ventilate the floor. This is metalled, as is the breast and roof of the mill body with iron sheeting.

Like Cross in Hand and Argos Hill post-mills, Jill had a large tailpole-mounted fan, the steps being carried on separate trolley wheels. It seems that this tackle was fitted at the time of her removal and replaced the top fan arrangement fitted at Brighton. The curbs for these are still to be found in the grass.

Jill, undoubtedly the lady of the pair, was very attractive when working, with the Sussex fan-tackle bringing her into the wind. Plate 34c shows her and Jack whilst still at work in 1905. Mr. Guy, the last miller here,[7] was in occupation and working for a Mr. Wood. Sadly, a little later, in 1908,[8] Jill was put out of action by a storm. She sustained considerable damage to her fan-tackle and I suspect a broken tailpole, for she was filled with chalk to prevent her swinging about. This was repaired very shortly after. Both Jill and Jack had stopped work in 1907, or a little before, when Mr. Wood found that business was difficult and would not support the upkeep of the two mills.

Jill subsequently deteriorated and was allowed to age, losing her fan-tackle completely and sweep details during the next 45 years, until, with the help of the Cuckfield Rural District Council in 1953/4, she was renovated by E. Hole and Son. The work included shoring up the breast with steel joisting.

Both of the mills here are now in the care of the Mid-Sussex District Council, and Jill stands permantly looking towards her former home, a little sadly perhaps, for Jack has turned his head away, and displays his new sweeps to the south-east.

I am pleased to add here that a full restoration programme has been worked out for Jill, and it is hoped that during the next five years, with the use of volunteer and professional labour, she will be brought to working order.

The proper way to approach this spot is to start from Clayton Farm at the bottom of the north scarp of the downs here and walk up. This allows us to see Clayton first with its part-Saxon church. The route here, however, is steep and rather muddy at times. The little road at the top of the hill makes the spot accessible by car. This lane runs eastwards off the B2036, Brighton to Burgess Hill road. [TQ 303-134]

CROSS IN HAND (New Mill)

Fatigued White Post-mill with Sussex Tailpole Fan-tackle.

This mill, owned by the Newnhams and worked by the Ashdowns, both part of the same old Sussex milling family, is what we might call a High Weald mill.

In company with Argos Hill and Nutley, she stands on the outer edges of Ashdown Forest, one-and-a-half miles west of Heathfield on the Lewes Road. In the vicinity, the Wealden Ridge is at its highest between the North and South Downs, the sandstone rocks being evident about here.

The mill itself stands above Cross in Hand on a perfect mill mound and at 531ft. is only a little below the height of Clayton mills, 20 miles south-west of here. This gives some idea of the rise in this sandy strata which starts well below the Downs.

Cross in Hand, like Clayton post-mill, has had an unsettled life. She started life at Mount Ephraim in the parish of Framfield. The age of the mill is a mystery. She is first noted on the 1840 tithe map[1] opposite the inn at Mount Ephraim, as owned by John Kenward and occupied by William. It is known that New Mill (as she was to become) was brought to Cross in Hand in 1855[2] by Medhurst for William Kenward. A mill is also known to have been erected at Framfield in 1806,[3] but this is not shown on the 1813 Ordnance Survey map or on Greenwood's map of 1823/4, and so it is not altogether certain that this became Cross in Hand New Mill.

The second home was a site about a quarter-on-a-mile south-west of here and the mill was not looked kindly upon by the local landowner, Squire Huth, who later decided to encourage its banishment from view. From there she was moved once again in 1868 to join a white hand-winded post-mill (which Mr. Kenward had owned since 1854) already here on the mound nearby.[4] The roundhouse of this can still be seen.

Today, just below the new mill, is a power-mill which was once run by a 39 h.p. producer gas engine. This incorporated stones and machinery once worked in the roundhouse lean-to and, like the mill house, was built some time after the arrival of the new mill. An older house named 'Mill House' still stands below the site of the old mill.

From 1882,[4] the present Mr. Ashdown's grandfather, Jabez, worked as grinder for Mr. Kenward's widow, and when she gave up in 1888, went into partnership with his first cousin, John Newnham (the present Mr. C. J. Newnham's father). They worked as Newnham and Ashdown. Mr. Ephraim Ovenden worked with John for a time before joining his brother, Thomas, at Dicker New Mill in 1913. John Ashdown succeeded Jabez in 1926[5] (Plate 37). On his death in 1937 the business became J. B. Newnham and Son, with Sidney Ashdown as miller. This continued up until 1969 when a stock broke whilst the mill was working. This ended the mill's commercial life and the Ashdown's 87 years of unbroken association with her. There were, however, earlier associations by both the Newnhams and the Ashdowns with the Cross in Hand mills.

A large feed company now has the business, but grinding ceased here altogether in 1971. The mill, however, still belongs to the Newnham family, and was the last to work commercially in Sussex.

Samuel Medhurst of Lewes, the 19th-century millwright, rebuilt this mill, using as much as possible of the original. The mill was fitted with Medhurst's tailpole tan-tackle at this time, replacing a top fan arrangement similar to that seen at Icklesham today. He had her erected on a 25ft.-diameter single storey of brick. The upper storey of the roundhouse is built of wood and protected with corrugated iron (Fig. V).

A 'spur gear' mill, she had three pairs of 4ft.-diameter stones. Originally, two pairs were under-driven in the head of the mill via a 4ft. Ø spur wheel under the stone floor which takes its drive from the 9ft. 8ins. diameter brake wheel. One pair of these was removed in 1933. The layshaft across the back of the brake wheel drives a kibbler in their place. The third pair are peaks and over-driven directly off the 8ft. 4ins. Ø tail wheel. The iron windshaft has a third bearing just behind the brakewheel to prevent undue vibration. (See p. 156.) The main work done here since the 1900s was in provender, for example the grinding of barley, oats and maize for feeds, etc. The windmill was latterly used for kibbling and grinding maize and the power-mill for grinding oats and barley.

The sweeps are of the patent type. The remaining pair and striking gear have now been removed for safety. A novel and very efficient invention of Mr. Ashdown's grandfather can be seen in the bin floor here. Millers working with the conventional striking gear, which loaded the shutters using the rack and weight wheel, found that on a day when the wind was variable, the friction in this system would hold the sweeps for a time after a heavy gust. Thus they would remain stuck for precious moments. Mr. Jabez Ashdown successfully overcame this. The weights were placed in a box inside the mill, which, supported by a lever assembly acted directly on the striking rod at the tail end of the iron windshaft. The weight box thus loaded with the correct weight would bob gently, as the wind rose and fell, keeping the sweeps set at the required pressure all the time (Plate 9). This system also has the advantage of the rocking lever method widely used in Kent, in that the latter, although an improvement on the weight wheel and chain, still has to be loaded outside the mill. An example of the Kentish rocking lever may be seen at Stone Cross tower-mill described later.

The size of Cross in Hand mill is perhaps best appreciated when standing at the bottom of the steps. From 18ft above, the tailpole strikes down and through the steps, some 12ins. square and 34ft. long to support the Sussex fan-tackle, which carries an 11ft.-diameter fan. This incidentally, was originally fitted with five blades by Medhurst and altered to eight by Neves in 1907. The tailpole has recently been replaced as it had rotted through and collapsed. The large fan-stage runs on iron wheels which have curious traction spurs in their rims. These wheels run in a gravel track unlike the steps, whose truck wheels run on an iron curb of their own (Plate 11).

The mill body is a large one, 21ft. from head to tail, 12ft. wide, and approximately 45ft. from the ground to the cap ridge. This is protected from the weather with iron sheets over the boarding, clad on breast, sides, and roof. Cross in Hand must be the biggest post-mill erected in Sussex with the exception, perhaps, of Windmill Hill, the rusting remains of which stand eight miles south-east of here.

The spout and stone floors in the body are surprisingly large and certainly not cramped to work in. The bin floor, however, is rather confined and houses three large bins, a sack hoist, plus the weight box assembly. The stone floor has two pairs of stones and the maize kibbler plus, of course, most of the brake wheel, tail wheel, and bins. The spout floor contains tentering gear for both pairs of stones and ahead of the post, the spur gear and stone nut for the head stones plus the meal spouts from the floor above.

The post, 2ft. 6ins. square at its base, was cut at Whiligh near Wadhurst[4] from a fine oak tree of great age. In the upper floor of the roundhouse which was used for storage, the other trestle timbers can also be seen. These are massive, which is not surprising when we consider the size of the mill they have supported for over 120 years. (Cast on the samson head bracket are the words 'S. Medhurst, Millwright, Lewes, June 1855'.) Also, we can see on the main post between the cross trees and quarter bars, Medhurst's mark '1868 M' painted at the time of its last move.

Neves, who had care of this and the power-mill in latter years, were employed to fit a good deal of iron strengthening to the mill in 1926,[6] particularly at the side girts and front corner posts. This has saved the main timbers to a great extent. The mill is now head-sick, the body having dropped and tipped forward a good deal over the years, and the whole structure is ageing (Plate 39). This may be the reason for the rather uncanny swaying sensation one experiences when up in the bin floor on a windy day.

Mr. Newnham and Mr. Ashdown both hope that one day this once fine Sussex post-mill will be fully restored, perhaps to work again. At present the fan is disabled and the sweeps unable to work with the wind. At the time of writing, negotiations are under way, and I understand that two fabricated steel stocks are to be fitted, also that some repair work on the breast is in hand.

To reach the mill we turn north off the A265 Uckfield/Heathfield road, up the mill lane in the village itself. [TQ 558-218]

HIGH SALVINGTON (Durrington)

Renovated Black Post-mill with Tailpole and Talthur

High Salvington is the first true West Sussex mill that we encounter. This, incidentally, was the only post-mill remaining on this side of the county before the boundary was moved east in 1974.

A mill is shown here in the parish of Durrington on Budgen's map of 1724, but it is not certain that this is the mill that stands today. High Salvington mill was insured against fire early in its life,[1] and is understood to have stood on the hill here since the 1740s.

A short history of the mill published recently by Worthing museum[2] tells us that a Mr. Daniel Redman was miller here from 1824 until 1847, when William Beard took it on. I believe that the family had the business from then until 1871 (during which time Charles Davey and Walter West had respectively been grinders for the Beards), when the following notice appeared in the *Sussex Advertiser* of 25 October 1871:

> To let with immediate possession a post mill with house, garden and appurtenances complete. Situate at Salvington near Worthing Sussex. For particulars apply to Messrs. Medhurst & Son Millwright Lewes.

The business was taken on by Walter Brown and worked until 1897 when the whole was purchased by Col. T. F. Wisden. A condition of sale ensured that the mill was kept in working order and it did in fact grind once more for a while in 1914. A Mr. Coote was the temporary miller.

An old postcard (Plate 41) of this mill shows her around the turn of the century with two spring sweeps and two commons. Two pairs of stones were fitted, one pair of burrs and one pair of peaks. The wooden windshaft, which had an iron canister end, drove the peaks in the breast via a clasp-iron brake wheel and the burrs in the tail from a fine well-braced compass-arm tail wheel. An auxiliary drive for a dresser and cleaning equipment was also taken from this wheel. The postcard also shows her standing in open surroundings with a sound wooden roundhouse almost unrecognisable as the mill that we see today. The present concrete roundhouse was built as a tearoom 67 years ago. Worthing Borough Council have cared for the mill since 1959, and employed Messrs. E. Hole and Son to renovate the body in 1961. This entailed completely refurbishing the weatherboarding, a new set of stocks, and four new dead sails.

When visiting here on one of the open days (Plate 40) I found the spout floor empty except for a pair of sack scales and a stone governor resting on the floor. Above, in the stone floor, the first thing noticed is the little window in the breast which looks out over the canister end of the windshaft. The view is out over Worthing from this hill, 300ft. or so above the sea. We can appreciate how easy it was for sailors of old to use this mill as a sea mark to aid coastal navigation. We find here that the brake wheel has gone, removed several years ago because of infestation with woodworm and death-watch beetle, two of the many deadly enemies of the windmill. The tail wheel remains, as does a pair of French burrs in the tail of the mill with its iron 'horse frame' and the peaks in the head. It is hoped that these and the remaining machinery will be restored to a working condition in the not-too-distant future.

On examining the trestle which is to be seen through the roundhouse windows, we find that it rests on ground-level piers. A point to be noted here is that, contrary to some popular belief, the post is not part of a growing tree, but, as are all conventional posts, clear of the ground. The difficulties involved in cutting those great mortices right through a growing tree can be imagined.

The steps to the spout floor were raised by a talthur pole in order to wind the mill, although when the mill retired, the tailpole was secured to a wooden post.

In March 1976 High Salvington post-mill suffered a broken sweep. The remaining three have been removed for safety. An inspection of the structure has revealed that she is badly weakened by rot and beetle, and now lists a little to the west. Sad to say, she is an unhappy site standing in long grass and surrounded by a safety fence. However, Worthing Town Council are in the process of setting up a £20,000 restoration project, led by Mr. John Pelling, which will involve the whole of the weatherboarding being stripped off and the structure de-infested. A set of diagonal internal bracing ties are also to be fitted and the body 'pulled' into shape once more. When this has been completed, sweeps will be re-hung and the mill restored to its former appearance.

This mill is found on a high hill above the back of Worthing, three miles inland, and is best reached by the road to High Salvington from the A24. A secondary reason, perhaps, for visiting this area may be seen a mile away across the Findon Valley. Six hundred feet above sea level, is the large earthwork known as Cissbury Ring with its Stone Age chalk ramparts.

[TQ 123-067]

HURSTMONCEUX (Herstmonceux) (Windmill Hill)

Derelict post-mill with Tailpole and Talthur

About a mile to the east of Hurstmonceux village, perhaps being once a wooded place belonging to a Sussex forbear, or as latterly suggested, named so from the marital combinations of Herst and Monceux or Monceaux as Arthur Mee's *King's England* has it, we reach Windmill Hill, a sharp rise in the land, although we are in fact already on a ridge overlooking Pevensey Levels. On the roadside stands this witness to an age past, and it is obvious that this mill has not worked for a very long time. In fact, it stopped in 1893 because of the unsafe nature of the weatherbeam and of the rest of the main structure of the three-floored body.[3] (Plate 44)

This, a giant of Sussex post-mills, approximately 49ft. from the ground to the roof ridge, was erected on this traditional site between 1814 and 1819.[1]

Akin to Cross in Hand and of very similar proportions, it seems likely that these two were built by the same millwright. Like many East Sussex post-mills the body was overclad with iron sheets; today it has the appearance of an aged knight in armour (Plate 42), and viewed from the next ridge inland, 'he', for I am sure that this is a male mill, seems to have paused on a life's journey along this road.

The iron canister points eastward. Here and there on the front and sides, the rusted cladding which was once painted white (Plate 41) has been disturbed to show the weatherboarding beneath. The back of the mill which now takes the brunt of the weather is unprotected.

In reality, this mill is very much rooted to the spot and has been for the last 160 years or so. The huge body stands on a tarred two-storey 22ft. 6ins. diameter roundhouse, inside the first floor of which we find the huge oak trestle raised 2ft. 2ins. above the actual floor. Here on the west side of the post is carved: TWB / HX / 1856 –probably the mark of one of the Beeney family who had the mill from 1845[2] till 1877. The Hammond brothers of Clayton took the mill in 1878[3] and owned it for 14 years. Mr. James Harmer actually worked the business[2] from 1878 until 1887; Mr. Hammond was overseer during this time. Mr. Henry Harmer and his family worked the mill from then onward, taking possession in 1892. It was during the first year of Henry's time that the mill was stopped and the large patent sweeps removed. Neves of Warbleton were employed to fit the roundhouse out to work from a steam-engine situated close by. Mr. Harmer was registered as steam miller from 1895[4] onward. Mr. Reg Buckmaster, who took over from the Harmers in 1969, owns the mill today and runs a thriving bakery business close to the mill and mill house. Mr. Buckmaster was kind enough to show me the original brick oven and the various other buildings used in earlier days, including a beer cellar which also provided yeast for baking. Several sties still remain behind the mill–these no doubt rang to the squeal of pigs, well fattened on the pollards and graded tailings from the dresser within the mill.

The whole, viewed from the opposite side of the road, is a very nostalgic site. I suspect that when the Hammonds took on this business, they had the mill clad in iron and raised from one to two storeys in order to improve its command of the wind and also had the ground floor extended outwards around its southern side which gives it an unusual two-tier appearance. (Plate 43.)

At this time the body is shored up inside with timbers and iron joisting jacked off the crown-tree in an attempt to counter the failure of the side girts. The front of the mill has also dropped under the weight of the wind-shaft, which is now packed off the weakened weather beam. As a result of this, the spout floor slopes up to the post a foot or more on all sides. Surprisingly, there is a fair amount of the original machinery in place, especially when we consider that the mill has not worked for 85 years. The iron windshaft still carries the wooden clasp-arm brake wheel and an 8ft. iron

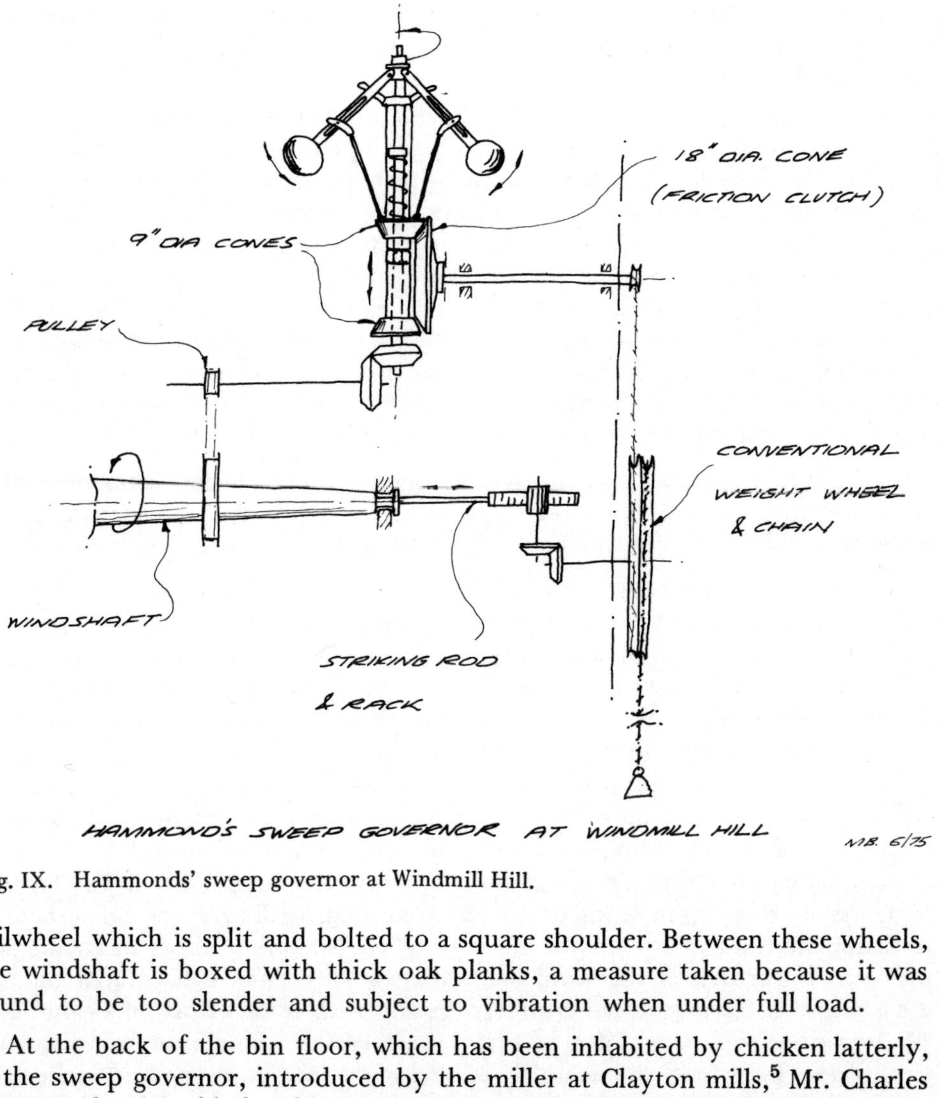

Fig. IX. Hammonds' sweep governor at Windmill Hill.

tailwheel which is split and bolted to a square shoulder. Between these wheels, the windshaft is boxed with thick oak planks, a measure taken because it was found to be too slender and subject to vibration when under full load.

At the back of the bin floor, which has been inhabited by chicken latterly, is the sweep governor, introduced by the miller at Clayton mills,[5] Mr. Charles Hammond, when his brother worked here. A belt from the tail of the windshaft drove this through bevel gears. The mechanism is basically a large centrifugal governor (Fig. IX), similar in action to the type used for tentering the stones. In this case, a sleeve with two opposing 9in.-diameter iron cones is raised or lowered according to the sweep speed. These in turn drove a larger clutch cone forwards or backwards, and via a light shackle chain and pulley, immediately pulled the weight wheel into the open or closed striking position. The adjustment of this system must have been very critical and it is interesting to note that the conventional weight chain was retained. In fact, as I understand the system, here and at Clayton tower-mill, the whole of the Cubitt's gear was retained intact and the governor incorporated to override the effect of the striking weight. When properly adjusted and favourable wind conditions prevailing, the shutters would automatically be kept at the required setting to maintain a constant working speed, even when the working load was altered within the mill.

Mr. Charles Edwin Hammond, of Clayton, who patented this arrangement in 1873, obviously intended it to be an additional refinement readily incorporated. The existing Cubitt's gear provided an accurate initial setting and also controlled the mill at the idle setting in the governor's action. A form of torque limiter was provided to ensure that the mechanism did not act too fiercely and strain the existing gear when increasing the sweep speed. The mechanism was used for a considerable time and in working order when the mill finished work in 1892. The basic concept of Mr. Hammond's patent was not confined to windmills, as a very similar governor and clutch mechanism was successfully used to raise or lower the pentrough gate automatically according to the waterwheel speed, admitting a greater or lesser amount of water power to the wheel.

It would take a very discerning windmiller to appreciate the claimed refinement and advantages of the sweep governor, and as far as I can ascertain, this example in the bin floor at Hurstmonceux post-mill is unique in windmills today.

In the floor below, two pairs of stones were overdriven, the 4ft. French burrs remain in the tail, but the peak head stones are amiss today. Against the tail wall, the dresser is still in place, the power take-off for this was by belt from a horizontal shaft, skew gear meshed with the back of the brake wheel. The dresser spouts hang in the spout floor, here as well the tailings were graded by a three-mesh separator. Other machinery found on this floor includes a grain cleaner and its blower, plus the tentering gear for the tail stones, which is rather refined and executed in iron. The bridge iron here carries Boaz Medhurst's nameplate, and controlling it is a hybrid type of weight control known as a Porter-loaded governor. This, and the similarity of some of the mill's features to Cross in Hand mill, leads us to conclude that Medhurst was concerned in re-fitting the mill at some time during the 1870s. Unlike Cross in Hand, this mill was brought into the wind by pulling on the yoke still evident on the great tailpole. Having turned Nutley post-mill with some effort on a calm day, 15 degrees or so, one wonders just how much labour would have been required to pull this huge bulk round into a freshening wind. As mentioned earlier, all post-mills are balanced as far as possible, and Medhurst fitted beautifully cast pintle iron and thrust bearings, but the inertia alone would have been daunting.

Within four miles of the sea, we are a mile from the world-famous Royal Greenwich Observatory, the grounds of Hurstmonceux Castle providing ample space for the many telescopes and laboratories here. The Observatory, of course, was until comparatively recently at Greenwich and moved to Sussex when the smoke-laden atmosphere of the city made observations impossible. The Greenwich Meridian 0deg. longitude originated from earlier siting and passes through Lewes, 16 miles west of here.

The site here at Windmill Hill is on the northern side of the A271.

[TQ 648-122]

ICKLESHAM (Hog Hill)

Renovated Black Post-mill with Top Fan and Trolley Steps

Here is a Sussex mill with a feature unique in the county. The winding fan is mounted above the roof (Plate 45) on two supports with stringers. A rectangular stage provides access from the back of the body. The drive is taken down the back of the mill to the right of the door by two interconnected rods, then via an early universal coupling to the bottom of the steps. These are now gone and have been replaced by a simple pole framing. Here, through bevel gears and a worm drive, the wind's action on the fan above arrived at the winding wheels and drove them round a track in the usual way. A convenient hand winder was also provided (Plate 14). At Cross in Hand, a double crank in the vertical drive from the fan serves the same purpose. This top fan arrangement was also fitted to several post-mills in Sussex, including Udimore and Winchelsea mill, which can be seen about 1½ miles away above the Brede Level to the north-east. Hog Hill mill, as it is called, after its present location, is only 1½ miles from the sea, about half-way along a very pretty lane which links Icklesham with Winchelsea. This lane passes through the historic south-eastern land gate of the latter village and reminds us of the feudal and fortified nature of the Sussex past.

A site immediately north of the church at Pett, a village two miles away, was the home of the mill until 189 years ago. Hog Hill mill, last shown at Pett on Yeakell and Gardner's map of 1785, was first shown at Icklesham on 1813 Ordnance Survey map. Sale notices in the *Sussex Weekly Advertiser* throw the required amount of light on the sequence of events.

An apparently new mill was advertised in the parish of Pett in December 1781: 'To be let and entered upon immediately; a good accustomed windmill'[1] This, and I take it be one and the same, would make the present mill 198 years of age!

On 31 May 1790 the windmill at Pett was advertised to be sold by auction on 11 June,[2] but the following week it was announced that the mill had been disposed of by private contract on 1 June.

In 1791, a miller, one William Sargent, insured his dwelling at Winchelsea for £200,[3] and the windmill at Icklesham for £400.[3] Subsequently, Hog Hill mill was worked by John Sargent from 1834[4] till 1855,[5] and then by Lewis Sargent until 1874.[5] The last to work the mill by wind were the Gardner brothers in the late 1920s. The Gardner family first came to Icklesham in 1890.[6]

The mill's tarred body has bin, stone and spout floors and worked two pairs of overdrift stones, one for flour and one for meal. The restored white fan (which has been mistakenly set to drive the mill out of the wind) and sweeps make an attractive contrast to the body and roundhouse which is two storeys high, brick-built up to the trestle and wooden above. Like Hurstmonceux, Hog Hill mill was stopped as a result of a weakness in the breast beam which supports the windshaft. The headstones were taken out in 1922[3] when this was shored up, and the side girts strengthened. The mill continued to work on a limited basis for a number of years, and it was then allowed to become rather dishevelled before being renovated by a millwright from Rye. The mill last worked with single-shuttered spring sweeps (Plate 46). These Thompson, the Lincolnshire millwright, replaced in 1968 with the basic frames that we see today. He also replaced the breast beam at this time.

On going over the mill recently, I found that the brakewheel has also been removed. The tailwheel, of wooden clasp arm construction, is still in place, mounted on a fine cast-iron box casting. A pair of 52in. French burrs remain in the tail and behind these, the dresser. The ground floor of the roundhouse, which is still kept well tarred outside, as is the whole mill, contains a full set of beam scales suspended from a joist. These were made by Jones at the Rother Iron Works, Rye.

Mr. Gardner used to run his business from the bakery and mill-house which stand close to the mill. These were combined in one building which was erected some time after the mill's arrival here. It is interesting to note the lengths that the builder went to, to integrate the house with the mill. He kept the two-storey building as low as possible, undercutting the mill mound to some degree, thus keeping the roof low on the seaward side. He also dug the footings on a radius struck from the centre of the roundhouse, giving its quaint curved layout (Platc 46). Again we find, as we often must in Sussex, that we are in a beautiful setting, with extensive views inland over the Brede Valley and southward across Pett Levels, to the sea.

This is a somewhat lonely spot, and whilst being perfect in summer, must be rather inaccessible in a bad winter.

Easily found to the south of Hastings to Rye road A259. **[TQ 888-160]**

KEYMER (Ditchling) (Oldland)

Renovated White Post-mill with Tailpole and Talthur

Finding this mill can be a problem, for it is not at all evident from the road, although it can be seen quite easily from the Downs above Ditchling and Clayton. A series of water-colours of Sussex mills by A. Foord Hughes[1] is of particular interest here, for the painter has captured the latter working days of Oldland mill's life, standing alone and unencumbered by buildings or trees here near Lodge Hill. Not so now, for immediately in front of the mill a number of 50-year-old oaks deflect the south-westerly winds and weather, once the miller's most reliable source of power (Plate 47).

The actual age of this mill (often claimed, like Icklesham and Halnaker, as the oldest in Sussex) is obscure.

An early mill was, according to Church records, worked in the parish of Ditchling in 1755.[2] Also, a mill is shown here on Yeakell and Gardner's map of 1783. After this, the evidence is confusing. Gream's map of 1795 omits the mill, the 1801 defence schedule[3] records 'Oldland Mill' in Ditchling, the 1813 Ordnance Survey map and Greenwood's map of 1823 omit her, but Pigot's *Directory* of 1828/9 shows Josh Beard as miller at Oldland mill, Ditchling/Keymer.

From this evidence, we can only make a guess or two. First, it is possible that over a period of nearly 30 years each of three map surveys mistakenly omitted the mill in turn and that she dates from at least 1755. Second, it is also possible, but less likely perhaps, that a succession of mills has stood here in the following sequence:

1. Early mill, gone by 1795.
2. Second mill built in time to be recorded in 1801, but gone by 1808 when the first Ordnance map was surveyed in this part of Sussex.
3. Third and present mill built in 1823/4 just too late to be recorded by Greenwood.

All later maps show her.

She was worked in early days by Mr. J. Beard, then from 1839[3] till 1858[4] by Joseph and William Winchester (Fig. VIII), and later by Mr. J. Turner till 1899,[5] when Mr. John White took her on. After his death in 1904, the business was worked by Mr. David Driver for his mother-in-law, Mrs. White. Mr. Driver was the last[6] to work the mill by wind in 1912, and when the First World War broke out, he left to work on his own account as a corn merchant in Hassocks.[7] (Plate 48.)

This attractive white mill became idle, and during the following years deteriorated until the last private owner, Mr. Turner, placed it in the care of the Sussex Archaeological Trust in 1927. During the following years, a number of appeals were launched, and in 1934 repairs were undertaken at the expense of a builder who wished to develop the vicinity, which, incidentally, he was not able to do. E. Hole and Sons were employed to carry out repairs at this time and a little later. The body had pitched forward slightly and was in need of strengthening and shoring-up.

Only two of the spring sweeps remain today, with springs and shutter bars in place (Plate 5), distinguishing this as the only mill in Sussex still to carry this mechanism in its original form (Plate 46). The body of the mill has been renovated once again in recent years and is shored-up through the roundhouse with timber and brick piers. It is, however, still leaning gracefully to the north-east. This was aggravated recently when the timbers were disturbed.

The tailpole and steps are in poor condition, and are, I believe, to be repaired shortly. Oldland mill was winded by the miller applying himself to a yoke on the tailpole. In its day, the mill worked two pairs of stones overdrift by wind (Plate 48). An engine in the granary building nearby was used latterly to drive the stones underdrift via a large pulley shaft and bevel gears (Plate 48). Inside the body, which has bin and stone floors with a large spout floor below, dressing machinery, etc., was driven by belt from a horizontal shaft. The stones remain today under the wooden clasp-arm brake and tail wheels, which are carried by a rather long and heavy iron windshaft, bearing the inscription B MEDHURST LEWES 1873. I suspect that Boaz Medhurst was responsible for fitting the auxiliary drive gear at this time, together with other iron work, including a fine sack hoist (Plate 35). Below the body, the original brick roundhouse is notable in that it is octagonal in shape, which is unique in Sussex and reminiscent of some Belgian post-mills. Within this, the fatigued trestle is borne by shallow brick piers. Sadly today the shattered brake no longer holds the sweeps, and the body is in need of renovation and girding up. Mr. Hawkins Byass, who lives in the Mill House, and the Hassocks Amenity Association are at present trying to raise the necessary money and enthusiasm to undertake a limited restoration programme.

Strictly speaking, this site is just on the Keymer side of the boundary between Ditchling and Keymer, and is locally known as Oldland Windmill. To reach it, we must travel a little way down the Keymer to Burgess Hill road to reach the mill lane. Even here, it is not always easy to see the mill, depending upon the amount of foliage on the trees. The lane (which is not accessible to cars) rises to Oldland and follows the greensand spur round into Ditchling. The mill is about half-way along this in the well-kept garden of 'Oldland Mill House', overlooking open land to the bottom of the Downs, two miles south.

[TQ 321-163]

LOWFIELD HEATH

Renovated White Post-mill with Tailpole and Talthur

The first illustration here (Plate 49) shows repair work being undertaken by E. Hole and Son in September 1971—when the mill was still within the Surrey boundary. As may be seen, a pathetically small sweep was still evident at the time, one of a set of four hung in 1939 when the owner, Mr. E. Lowes, had this once white mill made good.

Standing on a two-storeyed tarred brick roundhouse, she is a tall mill, this feature being accentuated somewhat by the narrow arching of the roof. Surprisingly, the body only contained two floors, stone and spout; the bins were placed above the stones. Also notable on comparison with the traditional post-mills of Sussex is the fact that the body is extended well back from the crown-tree. All post-mills tended to be built in this way to counterbalance the sweep assembly. In the case of Lowfield Heath, however, the millwright exaggerated this, placing only one-third of the structure in front of the post. This has certainly prevented her becoming head sick, a common failing in an old post-mill.

The history of the mill is a little confused. Researches reveal that it was popularly held that this was brought from Hookwood, 2½ miles north, in 1762, to replace an earlier mill.[1] The evidence available rather contradicts this, for it was not until 1820 that a windmill was recorded at Hookwood,[1] and similarly there is no evidence of the Lowfield Heath windmill before 1762.[1] A Mr. James Constable is said to have been the miller at the time, and the family appear to have owned the mill up until 1831.[2] Then came John Robinson till 1862,[3] followed by Henry and Joseph Robinson. From 1848 or so, Mr. John Ansell was the miller, and worked for Henry Robinson.

A note dated 16 January 1915 in the Gurney Wilson collection relates Mr. Ansell's recollection of work here. When he came, the mill was working two large pairs of overdrift stones, a pair of 4ft. 6ins. ϕ peaks and a pair of 4ft. 4ins. ϕ French burrs, driven by common sweeps. Two of the latter were badly damaged in a storm some years later and a new set of long patent sweeps were put up. The old square iron windshaft which carried a wooden brake wheel and an iron tailwheel (Plate 49) was drilled through for the striking rod by a millwright named Morley, who, incidentally, is thought to have erected Beech Farm smock-mill at Cuckfield in 1873. The work was carried out with the two remaining sweeps in motion and reputed to have taken a fortnight. By 1880, Lowfield Heath mill had ceased working by wind and was allowed to decline (Plate 50). The remains of the sweeps were taken off and subsequent owners did little to sustain her. Most of the machinery was still intact when Mr. Lowes became the owner, but after the Second World War the whole structure was allowed to become weatherbeaten and vandalised to the extent that by 1968 she was completely open to the weather and with her windshaft lying with its tail end on the stone floor. Fortunately, the proximity of London's second airport at Gatwick highlighted the plight of this landmark, placed as it is on this plain-like area of Surrey. Aircraft noise was popularly blamed for some of the damage, and it was decided, to the relief of the owner, that the structure should be repaired. E. Hole and Son, the millwrights, were employed by the Surrey County Council and the Society for the Protection of Ancient Buildings to make good the structure.

The whole of the Gatwick airport area, including Lowfield Heath, was incorporated in the 'revised' Sussex from April 1974. However, I am sure that Surrey folk will still 'traditionally' regard this as their own.

To the west of the A23 just before the road swings east to bypass the airport. [TQ 270-398]

MAYFIELD (Argos Hill)

Preserved White Post-mill (Red Roof) with Sussex Tailpole Fan-tackle

Almost due north of Cross in Hand rises Argos Hill, merely a hummock, yet at 605ft., it is one of the highest points for miles around. This, an old-established mill site, mentioned as far back as 1656 in this context when Nicholas Puxte of Garden House held 'Ye windmill field att Argatts Hill'.[1] As we approach the hill from the south along a wooded road from Five Ashes, the mill appears before us as the road becomes steeper and stands a short way along a lane to the right near the top of the hill. Here, we find this pretty lady, perched on a single-storey tarred roundhouse, a contrast to the towering derelict at Hurstmonceux. This mill, recently renovated, has a painted brick-red roof which is clad in iron and a little reminiscent of a Norman helmet when viewed obliquely (Plate 51). The whole is a credit to the Wealden District Council who care for it and Mr. Frank Child, who has had a long association and affection for the mill.

Understood to date from 1835,[2] this attractive post-mill was worked by a well-established local milling family, the Westons, continuously from approximately that time till 1912/13, when Mr. G. Wickens purchased her. Mr. Aaron Weston is first mentioned here on a tithe apportionment map[4] of 1844. Later, she passed to Captain Hardy and was last rented by William Richardson, an uncle to Neves the Heathfield millwrights. Neves' work on this mill during this time is shown in Chapter 8. After the mill finished for good, Mr. J. Fuller rented the ground for farming and became the owner in 1941. By 1955, when the Uckfield District Council took on the upkeep, the mill had deteriorated somewhat, and William Sands, associated with Neves, were employed to fit a new breast beam and carry out repairs to the side girts and corner posts, preventing collapse. The fine shutterless sweeps were Sands' work, but have since been replaced by E. Hole and Son of Burgess Hill, as was the fan-tackle in 1969. Holes have made a total of seven sweeps and four stocks for this mill, the last one and a sweep in early 1978 after storm damage here in October 1976. This stock is made of steel plate, fabricated by Holes at their yard in Burgess Hill and should withstand anything.

Something which is noticeable at once is the extended body (Plate 51). This must have been altered after the mill was built, for she is unbalanced, the extension housing having put the centre of gravity behind the crown tree and post. This is evident to some degree in that she leans backwards slightly. Because of this, no doubt, a set of truck wheels was fitted where the bottom of the spout floor bears against the main post, making her easier to wind.

1. The open trestle post-mill at Nutley.

2. (*top*) The post horns, Ifield post-mill.
3. (*bottom*) A cross-tree joint, Ifield post-mill.

5. (*top*) Detail of spring control on the remaining sweeps of Oldlands post-mill, Keymer.
6. (*bottom*) Patent sweeps at Shipley smock-mill.

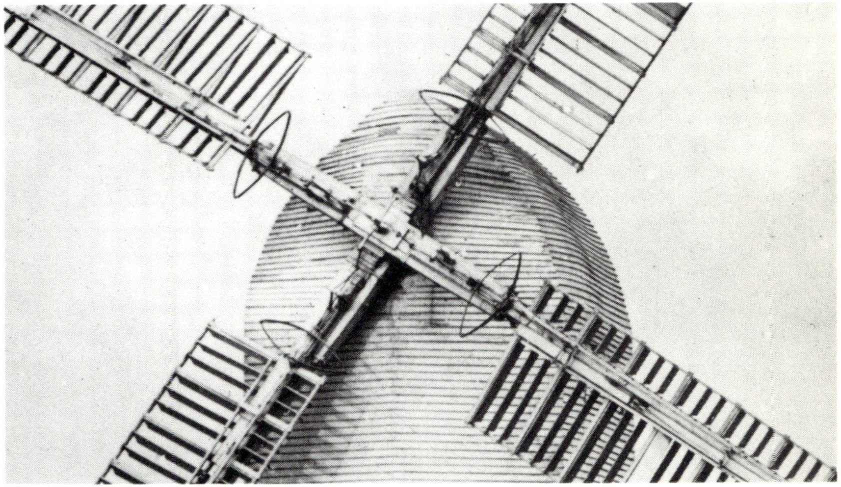

7. (*left*) Detail of shutters at Shipley.
8. (*top right*) Detail of Cheale's spring patents at Southern mill, Lewes, 1890.
9. (*bottom right*) Mr. Ashdown's weight box at Cross in Hand post-mill.

10. (*top left*) The tailpole and talthur of Nutley post-mill.

11. (*bottom left*) Winding wheel assembly: Cross in Hand post-mill in working order.

12. (*right*) Sussex tail-pole fantackle: Argos Hill post-mill.

13. Roof-mounted fan at Icklesham post-mill.

14. (*top*) Winding wheel assembly at Icklesham post-mill. Note the hand winder.

15. (*bottom*) The fanstage at Patcham tower-mill.

16. The winding worm and wheel from Littleworth smock-mill; the rack in the background.

17. The wooden clasp-arm brake wheel at Nutley post-mill.

18. The iron brake wheel at West Blatchington smock-mill–bolted, rim to spokes and mounted on a square shoulder.

19. Holloway's screw-down brake at Patcham tower-mill.

20. The brake lever at Cross in Hand post-mill.

21. The composite great spur wheel at West Blatchington smock-mill.

22. The iron great spur wheel and auxiliary bevel gear drive at Barnham tower-mill (underdrift).

23. A quant and iron stone nut for the tailstones at Cross in Hand post-mill (overdrift).

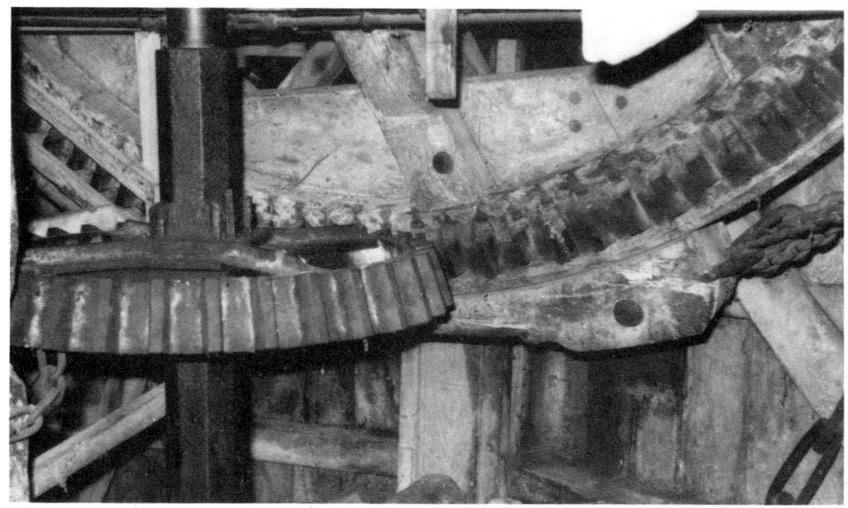

25. Cross in Hand. The wallower and brake wheel (spur-gear headstones)

26. *(top left)* Cross in Hand. The spur wheel and a stone nut for underdrift head stones.
27. *(bottom left)* Patcham tower-mill. Wallower and brake wheel.
28. *(right)* Stone drive at Barnham tower-mill showing the tentering, the footstep bearing, the jack ring and the iron stone nut engaging the wooden cogs of the great spur wheel (underdrift).

29. The peak stone casing and feed assembly at Shipley smock-mill.

30. French burr bedstone and peak runner stone from Bexhill Down post-mill.

31. Stone tentering gear at Stone Cross tower-mill.

32. (*top left*) The sack hoist drive at Argos Hill post-mill.
33. (*top right*) Auxiliary machinery; the smutter at Shipley smock-mill.

34. (*bottom left*) The dresser at Stone Cross tower-mill.
35. (*bottom right*) Sack scales at Cross in Hand post-mill.

36. Jack and Jill, Clayton, when working in 1905.

37. Cross in Hand post-mill in 1935, plus three generations: the first, Mr. J. Newnham, right; the Ashdown's son, John, left; and grandson, Mr. Sidney Ashdown, centre.

38. Jill post-mill, Clayton from the north, 1973.

39. Cross in Hand. The ailing post-mill, in 1974.

40. High Salvington post-mill. The renovated mill. April 1974.

41. A postcard of the mill, *c.* 1890.

42. On Windmill Hill, an ancient mill site.

43. Showing the unusual extended two storey roundhouse.

44. *c.*1870.

HURSTMONCEUX POST MILL

45. Icklesham post-mill. As preserved, June 1973.

46. *c.*1930, with her full complement of spring sweeps.

47. Oldland post-mill, Keymer, March 1974.

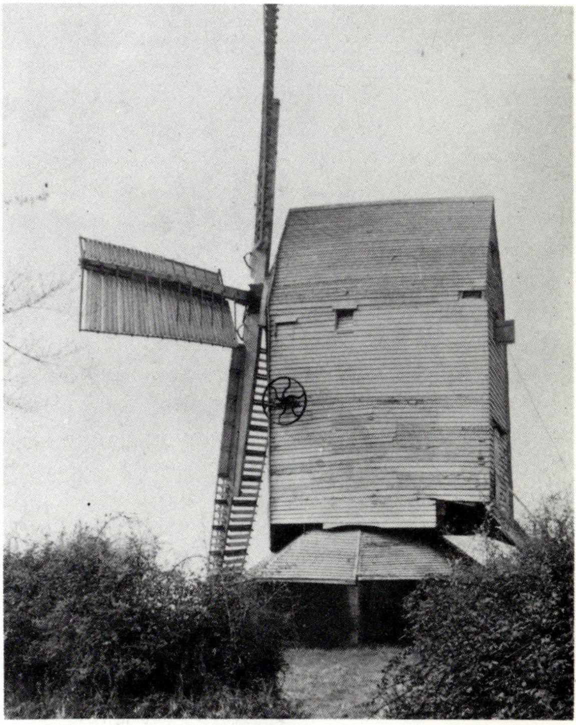

48. *c.*1914.

49. Lowfield Heath post-mill. During restoration by E. Hole and Son, September 1971.

50. *c.*1895.

51. Argos Hill post-mill, Mayfield, April 1976.

52. (*above*) Nutley post-mill as restored, March 1975.

53. (*right*) *c.* 1929.

54. Battle smock-mill, June 1976, displaying new aluminium sweep frames and fan.

55. Chailey smock-mill. The Heritage mill, March 1975.

56. As worked with large patent sweeps. Being repaired by a millwright, *c.* 1910.

57. Earnley smock-mill, August 1974.

58. Working in 1931.

59. Punnett's smock-mill. Archie Dallaway's mill, September 1974.

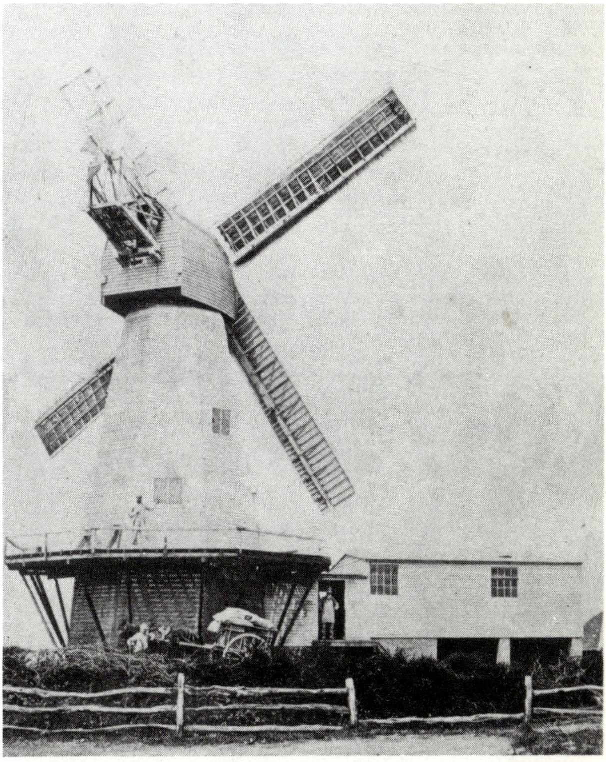

60. Cherry Clack mill as she stood in 1875: great-grandfather Samuel Dallaway in the doorway.

61. (*above*) Rottingdean smock-mill. The renovated shell on the cliffs, October 1975.
62. (*right*) Rye smock-mill, working *c.* 1912.

63. Rye, the replica, May 1974.

64. Shipley smock-mill. The well-preserved mill, July 1974.

65. West Blatchington smock-mill August 1970.

66. The rural scene at West Blatchington Farm, *c.* 1890.

67. West Chiltington smock-mill. The converted mill, February 1977.

68. The mill working approximately 100 years ago.

69. Arundel tower-mill, April 1977.

70. *c.*1890 (before the restoration of the castle).

71. Jack and Jill, Clayton, in working order, 1905. (The roundhouse of Dungate post-mill adjacent to Jack's tower).

72. (*left*) Clayton tower-mill: displaying new sweep-frames, March 1974.
73. (*above*) Minna Anson's chapel in the stone floor.

74. Barnham. The mill today. Part of the feed business.

75. After the power-mill took over in 1930.

76. Halnaker tower-mill, standing proudly on Halnaker Hill, August 1974.

77. The derelict mill, *c.* 1918.

78. Nutbourne tower-mill: a sad sight, July 1973.

79. *c.*1900.

80. Nyetimber (Pagham) tower-mill, August 1974.

81. *c.*1905.

82. Polegate tower-mill as restored, August 1971.

83. When working, *c.*1925.

84. Patcham tower-mill. Part of the Downland scene, July 1976.

85. c.1912. Note the spring spider.

86. Selsey tower-mill after conversion in 1908. (Holloway's men on the cap).

87. Selsey tower-mill. Part of a holiday site, August 1974.

88. Stone Cross tower-mill, May 1977. Note Mr. Hall's new fanstage.

89. Working in 1923.

90. (*top left*) Angmering post-mill working, *c.*1905.
91. (*top right*) Angmering post-mill. The roundhouse remain, March 1976.
92. (*bottom left*) Appledram post-mill. The roundhouse forming part of a residence, March 1976.
93. (*bottom right*) Ashurst post-mill. The ailing structure, *c.*1925.

94. (*top left*) Ashurst. The top of the centre post and pintle bearing (a garden ornament today), April 1976.
95. (*bottom left*) Bexhill Down post-mill. The remains of the trestle and round-house, May 1974.
96. (*right*) Bexhill Down post-mill: in working condition, *c.* 1935.

97. Cross in Hand old mill. The roundhouse, the new mill and power-mill.

98. Dicker (Golden Cross) post-mill, the two-storey roundhouse today, part of a residence, February 1975.

99. Dicker (Golden Cross) post-mill, *c.* 1904.

100. Lower Dicker post-mill, *c.*1902.

101. Lower Dicker post-mill. The roundhouse, used as a garage, April 1976.

)2. Harebeating. Today the two-storey roundhouse remains, with the post and shattered crown tree.

103. Harebeating post-mill, when working, *c.*1910.

104. Kingston (Ashcombe) six-sweep post-mill, *c.*1870.

105. Lewes–Malling post-mill, on a still winter's day, *c.* 1905.

106. Malling. The converted roundhouse, April 1975.

107. Northiam. All that remains of the brick piers, May 1976.

108. Northiam post-mill, *c.*1912.

109. Ringmer post-mill, when working, *c.*1910.

110. Ringmer. The centre post, September 1974, and the Weald towards Ashdown Forest.

111. (*top left*) Riseden post-mill, not long before the collapse, *c.* 1909.
112. (*top right*) The remains of the roundhouse, October 1975.
113. (*bottom left*) Sullington, the 'compass arm' windshaft, March 1976.
114. (*bottom right*) Sullington post-mill, in a renovated condition in 1910.

115. Winchelsea. The renovated body, May 1976, in its attractive setting.

116. Winchelsea post-mill, shortly after finishing work, *c.* 1900.

117. Baldslow smock-mill in full working order, *c.*1907.

118. Baldslow smock-mill, the converted remains, today part of a residence.

119. Billingshurst smock-mill. The base almost engulfed by bushes, July 1974.

120. Chailey South smock-mill: the remains of the base, May 1974.

121. Climping (or Clymping) smock-mill, *c.*1875.

123. Cowbeech smock-mill, *c.*1900.

122. Climping smock-mill. The base and part of the smock-tower integrated with the

124. Cowbeech smock-mill. The converted base, March 1976.

125. Cuckfield (Beech Farm) smock-mill: becoming redundant, *c.* 1910. A sweep is missing.

126. Today the base is still used as a store, April 1976.

127. Dallington smock-mill. One-eighth of the base wall (Mr. Dallaway's smock-mill at Punnett's Town is just seen on the skyline).

128. Dallington smock-mill with sweeps turning, *c.* 1906.

129. Hailsham—Hamlin's smock-mill, *c.* 1918
plus Rufus Tingley, grinder, and Mr. E. Catt.

130. Hunston smock-mill, *c.* 1916.

131. Hunston: the remains of the 10-sided base, February 1975.

132. Kingston smock-mill, showing the entrance to the original cellar base, April 1976.

133. Lewes smock-mill. The converted base of Town mill in Pipe Passage.

134. Now a tack store at Mr. Smyth's Lewes stables, the three-storey base of Shelly's mill, April 1976.

135. Lewes smock-mill; at the prison site; working as Shelly's mill *c.* 1870.

136. Littleworth smock mill, in full working order, c.1900.

137. The cut down base used as a stable, July 1974.

138. The remains of the machinery today at County Oak (now in Sussex),

139. Lunsford's Cross smock-mill, just after being tail-winded in 1905.

140. Lunsford's Cross smock-mill: the two-storey brick base, March 1976.

141. Ore White smock-mill: the base used as part of a garage business, April 1976.

142. Ore White mill, *c.* 1876.

144. Ore Black smock-mill: the converted square base, April 1976.

145. Staplecross. The converted square base, May 1974.

.46. Staple Cross smock-mill, in working order c.1910.

147. Warbleton (Chapman's Town) smock-mill, March 1976. Cut down 90 years ago and on the point of collapse.

148. Washington smock-mill, c.1918, on the open common.

149. Washington smock-mill. Once the home of composer John Ireland, the whole of the smock-tower was converted, c.1919.

150. (*top left*) Wisborough Green smock-mill, *c.*1900, a little after she last worked by wind.
151. (*above*) Topped with a new short smock, March 1976, the two-storey sandstone base converted into a residence.

152. (*bottom left*) Alfriston tower-mill, the tower forming the eastern abutment of the residence here, February 1975.

153. Alfriston tower-mill. The attractive setting for the mill here about the time that she finally stopped work 1906, from a postcard by Mr. Fred Judge.

154. Angmering, Highdown New mill in working order *c.*1870. (Note the similarity to Arundel tower-mill).

155. Angmering tower-mill, recently renovated from an empty shell to form part of the bungalow here, December 1974.

156. [illegible] tower-mill when working, c.1890.

158. East Wittering tower-mill. Standing in the spacious mill yard, *c.*1874. (Note the travelling stage for setting the sweeps).

159. East Wittering tower-mill. The ivy-covered brick tower in August 1974 (before the fire).

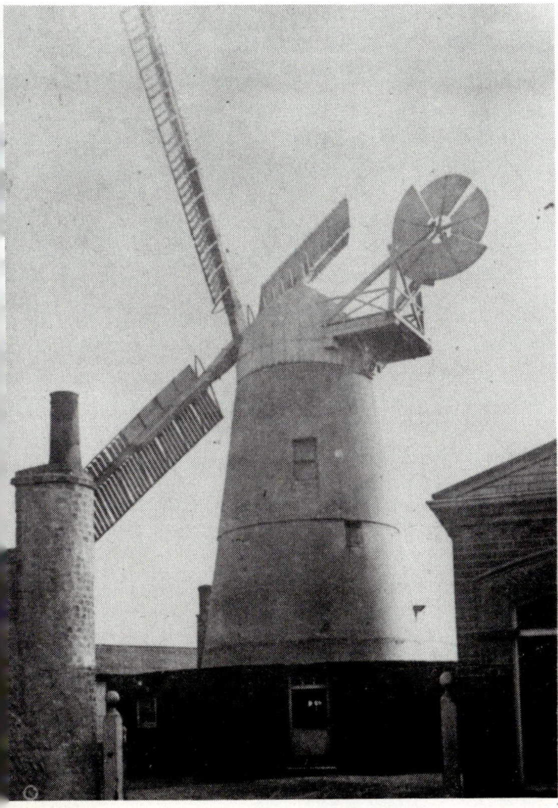

160. Eastbourne (Hurst's) tower-mill. The mill when working, *c.*1912.

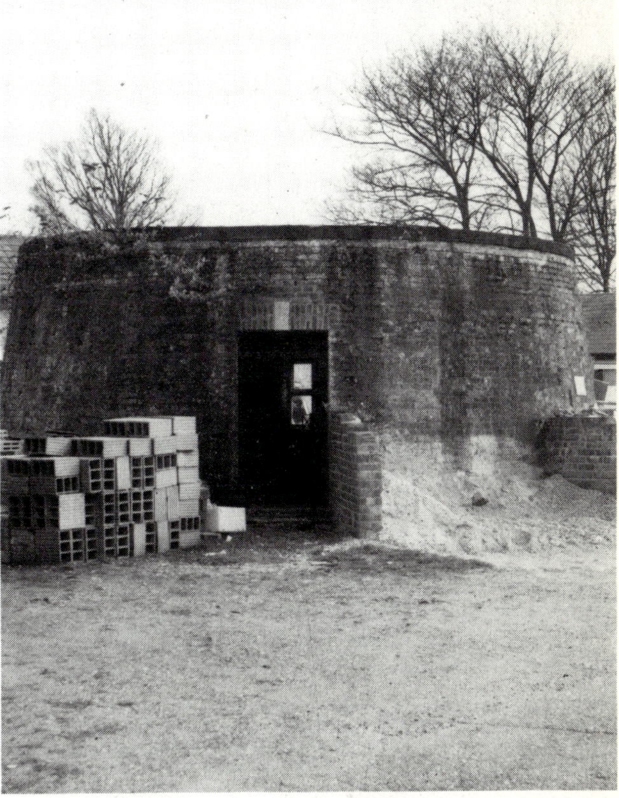

161. The ground floor of the tower now used as a store by Y. J. Lovell, Ltd.

162. Mark Cross tower-mill, *c.* 1909.

163. Mark Cross tower-mill. The remaining brick tower and power-mill building well utilised as part of a modern home, April 1976.

164. Worthing, Navarino mills, *c.* 1895, the south a tower-mill (nearest), the north a smock-mill.

165. Eastbourne (Pashley Down). The remains of a horizontal mill.

166. Ditchling, Elphick's mill. All that remains of the base, April 1976.

167. Ditchling, Elphick's mill in working order, *c.*1895.

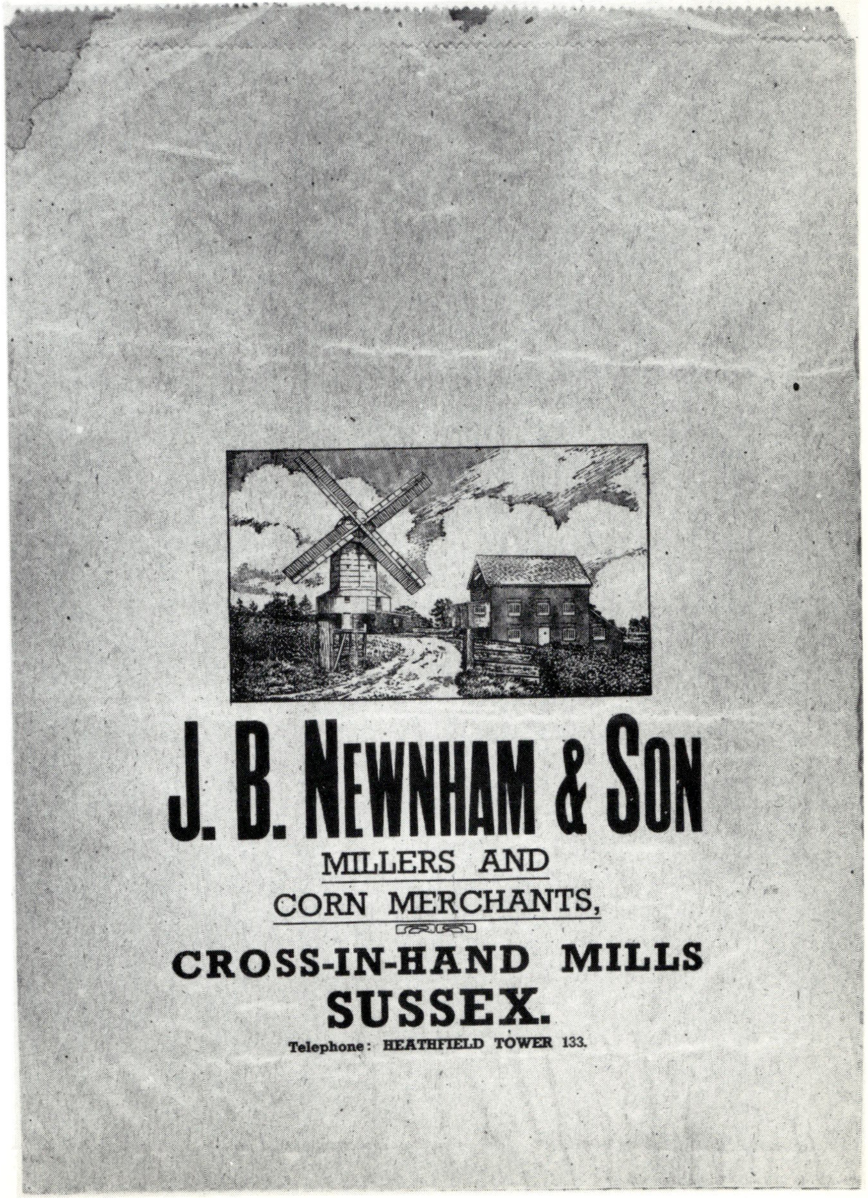

168. Paper flour bag from Cross in Hand, *c.*1938.

169. Trade calendar for 1904 from Crowborough tower-mill.

170. Mr. Sidney Ashdown dressing the stones within Cross in Hand post-mill, 1962, from Mr. F. W. Gregory's collection.

171. A selection of his dressing tools: a modern iron thrift, and mill bills; a traditional wooden thrift, and picks; the rubbing stone, and miller's spectacles.

172. The tool for centering the bridge, a jack stick, a miller's oil lamp, and the wheat staff for proving the level of the working face of the stone.

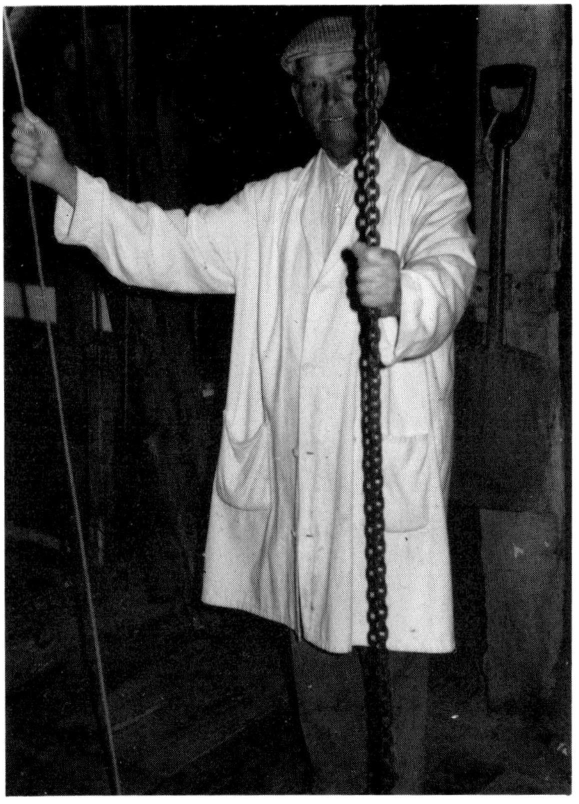

173. Mr. Gilbert Catt, the one-time owner of Hamlin's smock-mill, Hailsham.

174. Mr. Archie Dallaway, a unique Sussex millwright.

175. Mr. Frank Gregory, a unique Sussex windmillian, left, with the author assisting during restoration work on Jill post-mill, Clayton, Jan. 1979.

176. Mr. Clifford Hoad, the owner, and one-time miller at Bexhill Down mill, still the head of the family business.

177. Mr. John Newnham, the owner of Cross in Hand post-mill.

178. The miller's tomb.

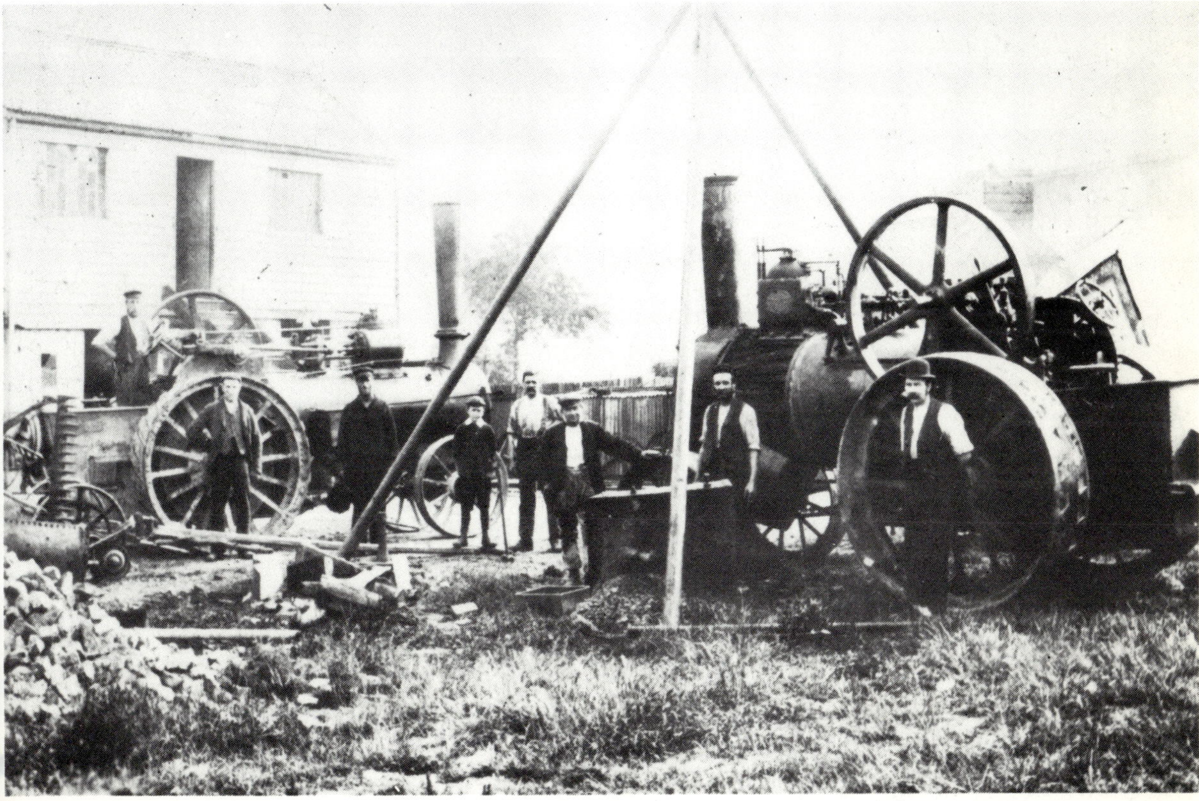

179. E. Hole and Son. The yard at Turnkey Lane, *c.* 1895; Ernest Hole, centre.

180. (*left*) Mr. Edwin Hole at the yard, June 1973.
181. (*above*) A new sweep for Shipley smock-mill, final assembly of the shutters, June 1976.

182. (*above*) The original brickworks forge.
183. (*right*) A wood and iron cart jack made for Mr. Ernest Hole in 1900.

184. Medhursts' millwrights' yards, in St. Anne's High Street, Lewes on the occasion of Samuel's 70th birthday in 1869. Samuel Medhurst and grand-daughter, centre.

185. The Medhurst iron gravemarkers in St. Anne's churchyard, Lewes.

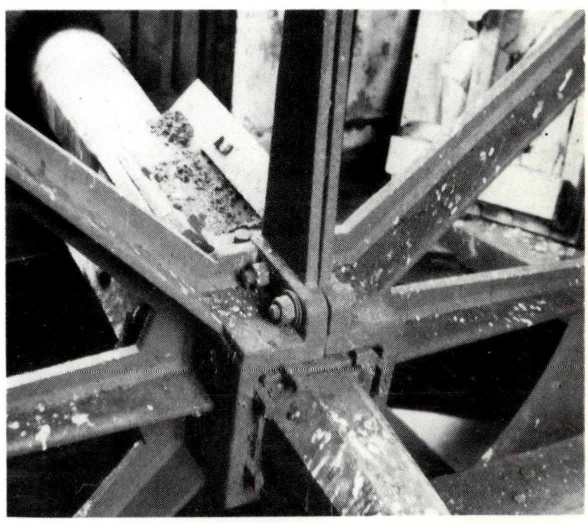

186. The cannibalised windshaft at Chailey, fitted by Neves in 1936.

187. The derelict remain of Glynde hollow post-mill, April 1977.

188. The fintail canister end of Mutton Hall post-mill, September 1977.

189. The reconstruction of Westham hollow post pumping-mill at Singleton Open Air Museum, April 1977.

190. Eccentrics and lift pump assembly.

191. Mr. R. J. de Little's working model mill at Broad Hill farm, Keymer, 1959.

192. As the model remains, February 1977.

193. Bishopstone tide-mills, *c.* 1880.

194. Birdham pumping-mill, 1934.

195. East Blatchington pumping-mill, 1937.

196. Saw smock-mill at Punnett's Town, *c.*1920.

197. (*above*) Wind- and water-mill at West Ashling. In a redundant state around the turn of the century.
198. (*right*) Standing in 1934 with most of the machinery still intact.

199. (*top left*) Angmering windmill wheel becoming derelict in 1931.
200. (*bottom left*) Alexandra Road wind-wheel, Heathfield, *c.*1910.
201. (*below*) Cornford's wind-wheel, Punnett's Town, *c.*1910.

The task of keeping the mill into the wind was performed by the fine Sussex tailpole fan-tackle which is very similar in design to that of Cross in Hand, and was likely a Medhurst addition. This had a smaller ladder up to the fan at one time. Two years after the mill stopped work in 1927,[5] the fan was blown off. This, paradoxically, has since been 'ventilated' to reduce its resistance to the wind, and the wooden cart-type winding wheels are pegged firmly to the ground.

When writing of Mayfield in *Highways and Byways in Sussex,* 1903, E. V. Lucas, on seeing the tracks for the winding and step wheels, made a rather amusing observation: 'Between Rotherfield and Mayfield is a little hill trim and conical–perfect in deportment for it has (as all little conical hills should have) a white windmill on its top. Around the mill is a circular track for carts, which runs nearer to the sails than any track I remember ever to have dared to walk on'.

Argos Hill mill was driven by patent sweeps, the shutters of which were removed in 1932. The weight wheel and box are seen still intact on the western side of the body; the striking gear, however, was removed in 1969. The sweeps today are non-working, though much more authentic in appearance than most dummy sails, with three stringers in each bay in place of the shutters to give an aesthetically pleasing effect.

In the roundhouse we find the well-restored trestle timbers. The main post, like that of Clayton post-mill, is built up from four lengths of timber. Today this is the home of a fine display of milling bygones and contemporary items. When visiting Argos Hill mill on one occasion, I was fortunate enough to meet Mr. Frank Child who is the curator. Mr. Child took time and trouble to show me over this very well-appointed post-mill. Each floor is neatly laid out. The spout floor is extended backwards to make room for the dresser spouts on one side and the miller's office on the other. Unexpectedly, the steps are convenient and those between the stone and bin floors pass upward through a rather cosy tunnel from the top of the extention. The ample bin space is very accessible, which is unusual in a post-mill. A high arched roof here allowed the miller to stand upright on a central floor to empty the grain into the bins at each side of him. Over two tons of grain could have been held here at one time (Plate 35).

Machinery inside the 20ft. by 10ft. 6ins. body includes two pairs of overdrift stones, one of peaks and one of French burrs driven by a 10ft. ϕ iron brake-wheel and on 8ft. ϕ wooden clasp-arm tailwheel respectively. A grain cleaner and a dresser, an oat-crusher and a whetstone in stone and spout floors were gear-driven from the circumference of the tailwheel. The dresser on the right, the other pieces by counter shafting from the left. It is interesting to find a mill that, having stood unused for nearly 50 years, still has as much working machinery, and for grinding wheat too. Other examples of this are at Shipley, a smock-mill, and the tower-mill at Stone Cross, both described later. Many mills, finding flour production by wind-power after the 1914–18 war uneconomical, went over to grinding oats and feeds and had the more refined equipment removed.

There are here two particularly unusual features. First, there remains in the spout floor, part of a hand stone-tentering arrangement. This is in the form of a multiple rope pulley system, and Mr. Ashdown, the last miller at Cross in Hand nearby, informs me is called a Spanish windlass. Second, the body door is made secure with a Sussex iron spring lock with its own patent key. This has been retained by Dr. Scott Williamson of the Mill Cottage.

At this pleasantly situated spot commanding extensive views over the East Sussex High Weald, we are in Sussex iron country, for it is around here at places like Rotherfield, Heathfield, Burwash and Wadhurst that the Iron Masters lived and worked in the 17th and 18th centuries. After this era, the industry moved away, as charcoal-smelted iron became less profitable than the northern counterpart, smelted in coal-fired furnaces. A great deal of our early iron came from the forests of Sussex. Hammer ponds and remains of workings still bear witness to these times. Cannon for the army of the day were cast from Sussex ore at foundries based around Buxted. There are still many examples of the domestic products of this industry in the great houses and manors of this county today.

The Wealden District Council, who open the mill regularly, have recently published a booklet about this mill called *Windmills and Argos Hill.* Several ponies occupy the mill field these days and seem more than happy to be in the company of this attractive lady. [TQ 571-283]

NUTLEY

Restored Black Post-mill with Tailpole and Talthur

This little lady represents the English post-mill in its basic early form, grinding corn for the local smallholders, with few of the refinements or trimmings of its later companions in this county.

From this phase in its evolution, the Sussex post-mill became the giant three-floored 'business' mill over clad with iron and standing above a two-storey roundhouse. The Sussex tailpole fan-tackle, latterly introduced, making winding automatic. Universally adopted patent sweeps were also fitted, relieving the miller of the tedious task of keeping the speed of the mill constant and correct. Our present subject remains charming in her innocence of such things.

Nutley, thought to have been brought here from Goudhurst in Kent[1] and perhaps genuinely the oldest post-mill standing in Sussex today, has defied all attempts to prove her so. From the appearance of basic simplicity and the apparent age of her post, brake and tail wheels, we might be correct in this assumption. The truth of the matter is that this mill is not recorded here before 1840. Inside the body, the inscription W.S. 1817[1] was to be found at one time, perhaps the mark of one of the Setford family who worked and were associated with this mill from its early days. One of the descendants asserted that his grandfather had the mill brought to this common land site from Crowborough. A Mr. Henry Setford is recorded at his mill here at Nutley on the tithe award of 1840.[2] Mr. James Martin followed in 1845, Mr. Luke Godley in 1851; then Mr. B. Martin from 1862 till 1874, when the last miller,[3] William Taylor took her on. He worked here until the mill finished work around 1908. The structure of the body had failed during the latter years of her working life.[4]

As we can see, Nutley is an open trestle mill (Plates 1 and 52) hand-winded by tailpole and talthur. The quarter bars and cross trees are open to all weathers, resting on brick piers buried in the mill mound. This, coupled with the attractive weathering of the breast to improve the airflow, gives this lady a rather top-heavy appearance. Until 1971, she was shored up on steel joists, and brick piers placed under the sheers of the spout floor. The owner, Lady Castle-Stuart, had these placed here to save the mill from collapse in 1928 when it was previously renovated (Plate 53). By 1972 the mill was working again for the first time in 64 years, due to a complete restoration of the trestle and of the main timbers of the body, including a side girt and the whole of the breast. The weatherboarding has been completely renewed with most of its framing and given several coats of creosote.

The whole mill is in working trim. This work (see page 143), which gained an Architectural Heritage Year Award in 1975, has been carried out by the Nutley Preservation Society. Working as a well-organised team, they brought it from a condition of near collapse in 1969 to the working condition seen today. A booklet called *Nutley Windmill Restoration* is available at the mill. This gives a full account of the work carried out here.

This mill is rather small, being 15ft. from head to tail and 10ft. wide. (Compare this with Cross in Hand at 21ft. by 12ft.) The centre post is very old and studded with hundreds of nails, perhaps to prevent the splintered grain snagging the miller's clothing. Beneath the spout floor, a huge wooden collar supports the structure to some extent. The 19th-century windshaft is of iron carrying 7ft. ϕ brake and tail wheels and replaced an earlier wooden one. This is evident in that both the wheels have been converted from compass arm to clasp arm. At the same time, no doubt, they were re-cogged to work with new iron stone nuts (Plate 17). When last working, she had a full complement of spring sweeps. Two of these are now canvas rigged as in earlier days, and have to be fitted with sail cloths to work with the wind. The setting of these, as mentioned earlier, requires the mill to be stopped every time a change is needed. To offset this, to some degree, the other two sweeps are still controlled by springs, which partially automate the system. These have been restored with new shutters and semi-eliptical springs.

The sweeps power two pairs of overdrift stones, the head stones are burrs, driven by the brake wheel, the tail stones are peaks and grind at times when the mill is working. This arrangement is unusual as burrs are usually fitted in the tail of a post-mill. There is no bin floor here, the stones apparently being fed by large hoppers forming a combined bin and stone floor as is usual in a small post-mill. A dresser is also fitted in the stone floor, driven by lay shaft from the tail wheel. The spout floor has the spouts from the stones and the dresser, also governors for the stones above.

We have come some eight miles west of our last visit, Argos Hill mill, and find ourselves almost in the centre of Ashdown Forest. The best way to find this mill is to walk from one of the picnic areas on the north side of the Nutley to Duddleswell road. This is almost the highest part of the forest and we have a grand view of this heathy expanse. To the east of here is Camp Hill Clump, believed to be the site of a Roman settlement, the masts of Crow's Nest, the Diplomatic Service Signal Station rise behind this. To the north is Gills Lap, and the high point above Crowborough Warren. We must walk westward following the fire break until we find ourselves on a shaded path. This leads past several cottages to the mill itself, which stands overlooking this part of the forest towards the village of Nutley. [TQ 451-291]

The next section is devoted to the remaining smock-mills of the county.

II

SMOCK-MILLS PROPER

BATTLE (Caldbec Hill) (King's Head Mill)

Converted White Smock-mill with Metalled Tower and a Fantail

This mill is part of a private residence, converted during the 1920s by Mr. A. N. Neve. Easily found, Caldbec Hill or King's Head mill, as it is known, stands as the name suggests on Caldbec Hill north of the historic village of Battle. I need not relate the story of the battle in question, suffice to say that this place was unknown before the fated Saxon King Harold met William of Normandy at Senlac Field here in 1066. The whole area hereabouts, from Pevensey Castle and Levels through Hastings to Winchelsea is steeped in real English history and should not be overlooked.

The history of this mill is often related and is well documented.[1] King's Head[2] mill was erected on the site of a post-mill which William Neve purchased from three farmers, Mr. Metcalf and others, on 29 September 1804. The post-mill was pulled down within a year and the present mill built. The Neve family worked the mill until 1839, when a Mr. Porter took on the business.[3]

Twenty-one years later, the whole was sold to the Battle Estate and remained so, being worked by Henry Harmer for a time before going to Hurstmonceux, and then by the Jenner family until the First World War,[4] when the mill fell into disuse. In 1924, Mr. A. N. Neve bought the mill and buldings. Mr. Neve had the machinery removed and the smock-tower with its single-storey brick base converted into living accommodation. Various working parts were used to enhance the adjoining house and garden. The brakewheel, for instance, was used as a circular seat, and a pair of stones as a sundial and hearthslab. The octagonal wooden smock was sheathed in zinc in 1894 to reduce the expense of weather-proofing and white paint. This has preserved the main fabric of the mill. The large hooded cap originally tarred black, was re-boarded with plastic in 1972, and the stocks and fanstage sprayed with fibre glass and resin (Plate 54) in order to save them from further decay. More recently, Mr. McNally the present owner, has restored this large smock-mill to something like its former noble appearance, albeit a little shiny. A new set of flat aluminium sweep frames which, paradoxically, if of working form would drive the mill backwards, have been fitted, and a metal-bladed fan set up on the stage (Plate 54). Further considerable expenditure is evident within the smock. Under the cap, where a motor is to be fitted to drive the sweeps, the bin floor has been converted into a cocktail bar. The stone and spout floor below are also luxuriously fitted out as a bedroom and lounge. Not surprising then that the mill, the house and the converted granary which stand in an extensive garden were offered for sale at around £200,000 in 1976.

When working, King's Head mill had 35ft. shuttered sweeps mounted on an iron windshaft. These drove three pairs of peaks.[1] The business was in provender, but I understand that wheat was rough ground in one pair of peaks, and then passed through a power-driven roller mill elsewhere on the site. The resulting meal was dressed and graded conventionally for flour production.

The original Sussex iron windshaft is today in the Mill Museum at Polegate. The sweeps are supported by a short dummy windshaft placed in the cap at the time of the conversion. This carries a large spur gear in place of the brake wheel. Like many smock-mills still standing, this has ended its days as living accommodation.

Easily seen when approaching Battle from the north, found to the west of the B2092. [TQ 748-166]

CHAILEY (Heritage Mill) (Beard's Mill)

Externally Renovated White Smock-mill with Fantail

From the South Downs Way, which passes five miles to the south, this rather slender lady can be seen standing at the approximate centre of the county. The mill and granary, which stands close by, were purchased from W. W. Grantham after a period of rental, and have been cared for since 1933 by Chailey Heritage.[1] This, a home for handicapped children, has units spread over Chailey Common, which is part of the western extreme of Ashdown Forest. Although not very high here, 246ft. to be exact, the surrounding area is open. The mill would have been able to take full advantage of the prevailing south-west wind coming over the Low Weald from the sea beyond the Downs (Plate 55).

Chailey smock-mill rightfully belongs to Highbrook, an attractive High Weald hamlet 5½ miles north of here. Known as Hammingden mill, she was erected on a site just to the south of the present church in approximately 1830.[2] The tithe map of 1841[3] shows Mr. John Arnold as working here. In 1844,[4] Mr. Bollen of Newhaven required a replacement for his mill which had burned down. He was able to purchase Hammingden mill and had her brought 19 miles south to a site overlooking the sea, and high above western Newhaven. As Newhaven smock-mill, she was no doubt useful, but by 1864[4] Messrs. Stone and Towner were the owners, and in the interest of good business, had just built a steam-mill at a harbour site near the famous 'Tipper' brewery. Mr. Beard of Chailey who at that time owned an old post-mill and required a replacement.[2] Hearing of the redundant state of Newhaven's smock-mill, he employed Medhurst to bring her back up to Chailey where she has resided for the last 115 years. A privileged mill indeed, for she has worked on Downland close to the sea after leaving a High Weald eyrie, and now stands in stately retirement on the lower flat common-land not far from the original home at Highbrook. Mr. Beard's millers were Mr. Godley, Thomas Comber, H. K. Walls, 1870, and A. G. Sparkes, 1874.[6]

The well-known East Sussex millwrights, Neves, knew this mill well and were responsible for rebuilding in 1933 after a devastating gale completely decapitated it in 1928. Once again, several years later, when a similar accident occurred, breaking the windshaft in two between the brake wheel and canister, (see page 159), the mill was restored by Neves to the state that we see today.

The second photograph shows Chailey about the time that she was last worked in 1911[5] by Mr. Lockyer, who had the business from 1878. The mill still boasted the original large set of patent sweeps, larger cap and steady pole (Plate 56). The illustrations show her today with somewhat smaller dummy sweeps and a skeletal fan. The windshaft that Neves finally placed in the cap in 1936 was the cannibalised shaft from Punnetts Town Cherry Clack mill, which was partially dismantled by Neves in 1935 (Plate 186). Today, a large pulley wheel is wedged on to this and bolted to the cap sheers. The yew tree close by is said to mark the centre of Sussex incidentally.

During its working life, the mill had three pairs of underdrift stones driven latterly by an auxiliary engine via an external pulley. The white smock-tower, now devoid of machinery save part of the upright shaft and the great spur wheel, had three floors: bin, stone and spout. This is held securely to the single-storey brick base by stout external tie-rods which are grouted into the ground itself. The whole is used today as a Scout headquarters for the Heritage.

From the fanstage the miller of the day would have seen the sweeps of Nutley, Argos Hill, Cross in Hand, and Jack and Jill at Clayton, weather permitting.

Just north of the Haywards Heath to Uckfield road A272, this still pretty lady is worth visiting whilst walking across the north of Chailey Common from the little road which serves the craft school here. [TQ 376–214]

EARNLEY (Somerley Mill)

Partially Restored Small White Smock-mill with Black Cap and Fantail

As with other windmills of the Manhood Peninsula and South West Sussex, Earnley was built with an attractive weatherboarded beehive cap. Those at Climping, Halnaker, Hunston, Selsey and East Wittering spring to mind. Immediate comparison may be made with Climping and the 10-sided smock-mill at Hunston, and we may conclude that these two and Earnley were built by the same millwright between 1799 and 1803. Whilst the exact date of this mill is not known, it is first referred to in 1803[1] and shown on the 1813 Ordnance Survey map.

When I first espied Earnley, or Somerley mill as she is sometimes known, she was rather dilapidated (Plate 57), the weatherboarding of the smock showing signs of rot. Above this, the fan had gone and the cap had a distinct list, perhaps due to decaying sheers. How long, I wondered, would it bear the weight of the windshaft and sweeps, a sentiment shared by the owner, Mr. C. N. Darby and his son, Peter, who have undertaken the start of a long and extensive restoration with the help of Mr. A. Pollard. Two sweeps remain, these are springs, the other pair which were commons, were sawn off some 13 years ago.

This mill was still working in the early part of the last war, producing provender. The Stevens Brothers in whose family the mill had been since 1845,[2] owned the whole business and their miller at the end was Mr. Ellis. On his retirement in 1942, Bartholemews of Chichester[3] took the mill on, as they had at Hunston. Sadly, the windmill was allowed to deteriorate from then until Mr. Darby came in 1958 (Plate 58).

When working, this mill had two pairs of overdrift stones, one of which was of the type known as composition (see page 162) which is a man-made version of burr stone, though rather harder. A third pair of peaks and dressing machinery were driven by a Hornsby oil-engine housed in the building which adjoins the mill. A Cornish-boiler steam engine was employed here earlier. The brick base is in good condition and now used as an outhouse. An inscription, FB 1827 was at one time evident in the brickwork. This, according to Mr. Stevens, commemorated the heightening of this mill by building a new base.

Being rather small, this mill has a dainty look, especially when we compare it with the rather forbidding and somewhat larger relation at West Chiltington. Earnley must be a lady, all be it slightly inebriated.

Tucked in the corner of a private residence, Somerley stands next to the B2198 road from Bracklesham and East Wittering to Birdham.

[SZ 817-984]

PUNNETTS TOWN (Blackdown mill) (Cherry Clack mill)

Rebuilt White Smock-mill with Fantail

This mill was a very pleasant surprise to me. On reading Hemming, I found that it had finished work by the late 1920s[1] and that the cap and machinery had been removed in 1935, the whole then demoted to a store. Later writings gave some hope, saying that the mill was partially restored and carried two sweeps, this, a condition that many mills find themselves in when a stock breaks or a sweep is smashed by a destructive storm. It is often mistakenly thought that the result is a two-sweep mill; not so. Recent Sussex mills were endowed with four sweeps only, with the exception of one breath-taking example, a six-sweep post-mill which stood above Kingston near Lewes.

Whilst on holiday eight years ago, I made a visit to Punnetts Town. Approaching from the south from Pevensey Levels through beautiful country-side full of fruit and ripening crops, the mill and its house appeared high above the surrounding terrain some five miles away. Passing into a valley below the village, I lost sight of her until almost upon the scene. Along a narrow lane some 580ft. above sea level, I came upon Cherry Clack mill, as she is affectionately known.

My surprise was to find that the mill had a full compliment of sweeps, though dead sails, now capable of carrying canvas, and that the mill will run, having been re-capped so to speak, and completely rebuilt internally (Plate 59).

Mr. Archie Dallaway (see page 141), the owner, is a remarkable man of many parts. He is a great-grandson of the original owner and a surviving member of a large old-established Sussex milling family.

A post-mill stood here originally and was known as Blackdown or Blackheath mill, after the site. This was burned to the ground in September 1859[2] (see page 158), and the question of a replacement pressing. It was decided to bring the smock-mill from Biddenden in Kent. Mr. David Hobden and Stephen Neve were employed to do this, for Mr. Samuel Dallaway[3] (see page 158). Cherry Clack mill was completely dismantled and removed to Punnetts Town. The rebuild is understood to have taken place in the same year, and the date 1859 is carved in the main roundel floor joist. This is charred at one end and appears to be the tailpole of the old post-mill. The photograph (Plate 60) is of particular interest here, for it shows old Samuel Dallaway a year before his death in 1876 with his three sons, Charles, Thomas, and John.[3] Samuel is seen standing in the granary doorway, Charles with the horse and cart laden with ground meal, Thomas on the stage, and John, who was Archie's frandfather, perched on the fanstage. After Samuel's demise, the three brothers continued as Dallaway Brothers for many years working this mill and the post-mill at Rock Hill, Burwash. Later in 1910 or so, John's son Demas took charge of the business[3] and continued until the finish of wind-working in 1929. The Dallaways worked this as a flour and grist mill, and also used its power to drive a saw-mill at one time. The smock-mill came to be known both as Blackdown mill and Punnetts Town Cherry Clack mill, 'Cherry Clack' being inherited from its former position in Kent amongst the cherry orchards. She was no doubt credited with the same bird-scaring properties as mechanical and often wind-driven 'clacks' which were used.

At the time of its rebuilding at Punnetts Town, the mill had a Kentish hood and a sloping fanstage very similar to that fitted at Stone Cross tower-mill some years later. The windshaft that Neves put up was a combination of the fintail canister end from the Blackdown post-mill and a new iron tail end. The old wooden end was presumably burned away in the fire. Plate 60 shows her working with two patent sweeps and two commons, and I understand that Mr. Archie Dallaway's grandfather, John, made a number of sweeps for this mill himself. Later on, the mill was fitted with four patent sweeps controlled by a Kentish lever system. An attractive water-colour showing this is to be found in *Windmills in Sussex*, by A. Foord Hughes.[4] When Cherry Clack mill finished work in 1929 due to a broken cap curb, she had three pairs of stones, two pairs of burrs on the stone floor, and one pair of peaks in the spout floor which could be run by an engine in the power-mill adjoining the mill. Saw-mill work was also carried out by a smock-mill which stood lower down the lane until 1933. This smaller mill, known as Punnetts Town Saw smock-mill, was brought here in 1866 from Horam and was the source of power for the saw benches of a timber yard here. When Cherry Clack mill was dismantled, all the machinery was removed, and most of it sold.

Imagine then, the task before the present Mr. Dallaway, when, in 1946, he decided to rebuild his great-grandfather's mill to a working standard. Since that time, he has completely renovated all the timbers of the octagonal smock-tower, including the seven-inch double-lapped weatherboarding. Here, some of the cant posts originate from the rebuild in 1859, and are large triangular baulks with 10-inch sides, and the others are of spliced 10-inch timbers repaired near the sill of the roundel, these being from the original Biddenden mill. Mr. Dallaway pointed out that some of the timbers were undoubtably taken from the original Blackdown post-mill as well, these being used mainly in the flooring.

Above the smock-tower, the cap and fanstage are a complete rebuild to Mr. Dallaway's own pattern. The 10ft. diagonal cap is of aluminium sheeting fitted over a wooden frame. Today, the fanstage is basically of a traditional Sussex design, this being served by a penthouse door. The six-bladed fan is somewhat smaller than usual for a smock-mill, at just over 6ft. in diameter overall. This drives the cap very smoothly and is fairly sensitive to a change of wind. The transmission here, built mainly from agricultural machinery components, is through a worm and wheel and reduction gears to a wooden lantern pinion engaging with the rack.

The iron windshaft was brought from Staplecross in 1951, when the mill there was dismantled (see page 141). This was drawn up through the stair-well and fitted through the iron centre of the brakewheel. The rim of the latter is 8ft. 4ins. diameter, built up of 16 oak feloes in two overlapping rings, interconnected with the spokes of the iron centre. The wooden driving cogs are set in mortices in the traditional manner. Meshing with these is the original iron wallower carried by an iron upright shaft, which, passing through the bin floor terminates with Mr. Dallaway's great spur wheel. Like the brake wheel, it is very craftsmanlike, made from sycamore and clasp-arm in configuration.

Here on the stone floor, we find three pairs of overdrift stones. Two pairs of modern emery-faced stones are housed in beautifully executed 'tuns' being round and built up in the same manner as a barrel and banded with iron hoops. One pair of burrs are housed in the more usual octagonal wooden stone casing. The drives here are built up of wooden stone nuts and fabricated iron quant or crutch poles. Below, on the spout floor, is an oat-crusher which Mr. Dallaway has installed, having rescued it from a local scrap yard. Above our heads here is the tentering gear for the three stones, part made from components rescued from less fortunate mills, notably Bodle Street and Chapman's Town. The two-storey base is built of local sandstone and 16ft. across corners inside with 2ft. 9ins. thick walls at ground floor level. Again, somewhat smaller than usual. A later brick extension surrounds this and has an iron railing around its roof forming the stage at spout-floor level.

Mr. Dallaway is now past the usual retiring age, although not completely retired from his vehicle repair business, which he carries on in the adjoining buildings, nor from his constant vigil against the elements and infestation in his mill. Most of the new timbers used here were locally grown, and often chosen whilst growing, and then felled. All the sawing and finishing was carried out on site, as was the carpentry and wheelwrighting. We must add to this fabrication of the many drives and fitting of parts recovered from other mills, to comprehend the extent of Mr. Dallaway's work.

When I first visited here a new pair of common sweeps had just been completed, 26ft. 6ins. long and 5ft. 3ins. wide. These were on trestles and freshly painted, standing in the old power-mill, which, driven by a 20 h.p. Petter horizontal oil engine, continued to grind feedstuffs for some time after the windmill ceased to work. Cherry Clack mill may not clack today, but is a very memorable tributed to Mr. Dallaway's skill, patience, and not least, his ingenuity.

Easily reached from the Heathfield road B2096 and seen from miles around standing high on this inland ridge. The picturesque mill commands a panorama of this lovely part of East Sussex with views into Kent and to the sea over Pevensey Levels and Eastbourne, the home of its younger cousin at Stone Cross, and the tower-mill at Polegate. Over to the north-west, we can see the red helm of Argos Hill post-mill, Mayfield. A lucky passerby is often privileged to see the sweeps joyfully swinging round in a light breeze as Mr. Dallaway airs his mill. [TQ 627-209]

ROTTINGDEAN (Beacon Hill)

Externally Renovated Black Smock-mill having had a Fantail

Rottingdean is situated on the brow of the Downs in a steep-sided cleft or gap, which once had easy access to a large shingled beach. The sea has greatly eroded the chalk cliffs, to such an extent, that a concrete walkway forming a breaker to the sea was erected early this century. This forms the undercliff walk, and continues on to Black Rock at Kemptown.

Beacon mill (Plate 61), as she is known, was erected on a single-storey brick base at the present site in 1802 for Mr. T. Beard.[1] An item in the *Sussex Weekly Advertiser* for 7 June 1802 describes the finding of a skeleton by workmen digging the foundations for Mr. Beard's new mill. The workmen left the site to go to dinner and discovered on returning that the remains and a burial sword that had been lying in the grave were missing, never to be seen again. The mill obviously has the honour of marking the last resting place of an ancient warrior.

The site here is a very good one for wind. Mr. Beard and subsequent millers no doubt took advantage of this. The smock-tower, however, suffered and required constant attention. Just how quickly the millers' wind became a destructive enemy is highlighted by the fact that Mr. George Nicholls, the last miller, finished work here in the early 1880s[2] and by 1890 Beacon mill was regarded so dishevelled as to require immediate demolition. Fortunately this was delayed and emergency measures taken. In 1905–6 the Marquis of Abergavenny, on whose land she stood, had the mill renovated to prevent further decay, but by 1922 she had become badly weatherbeaten once more. The Reverend Lewis Verey established a fund for restoration, and purchased a 99-year lease. Brighton Corporation have since become the landlords.[3] It was not until 1935, however, that enough money was raised to make a complete restoration. By this time the mill had lost her patent sweeps, the fan, a good deal of weatherboarding, and machinery. Mr. J. Yapp, of Haywards Heath contributed the balance required by the fund, and by August 1936 the mill was trim again. The weatherboarding was contracted to a builder from Broad Oak, various millwrighting to F. Neve and Son of Heathfield, and the sweeps to Holman of Canterbury, the well-known Kentish millwright.

It will be noticed that the mill has been blown backwards by the sea winds, and has a permanent set to the north-east. This was bolstered in recent years by an internal bracing frame of steel joists erected by E. Hole and Son,[4] our present-day Sussex millwrights, who also fitted new sweeps (Plate 61). The whole of the outer fabric was again completely renovated by Margetts and Cooke of Brighton in 1975.[4]

Here we have a Sussex smock mill typical of its time, with broad hooded cap and tarred body, standing on a tarred brick base, working three pairs of stones underdrift, and like West Blatchington, West Chiltington and Icklesham mills, has a slightly sinister air. We can, perhaps, imagine these being associated in times past with smuggling and the like.

Under the back of the cap, the iron winding worm is still in place. This was driven by a stage-mounted fan plus a hand-winding wheel. There must also have been a weight chain for controlling the striking gear of the four patent sweeps, but all of these, like most of the machinery inside, are long gone. The upright shaft with its great spur wheel is, however, still in place.

A bonny mill, and still a fine seamark, this can be reached by a short, steep walk from the back of the village proper. From a corner in the old High Street, we may climb the western side of the gap, past the village allotments, up to the edge of the golf course, which the mill will be found viewing in her retirement. A most attractive sight with the sea as a back drop, we cannot miss this mill.

A Rottingdean mill is said to have been used as a model for the motif used by W. H. Heinemann, the well-known publishers. This, however, is a stylised symbol showing a post-mill, perhaps after one that stood in earlier times on the eastern side of the village on a plot of land known as Mill Laine.

[TQ 366-025]

RYE (Barry's or Gibbet Mill)

Poor Reconstruction of a White Smock-mill with Fantail

This has been the subject of many paintings, for the mill is pleasantly situated on the eastern bank of the River Tillingham as it enters Rye town; unfortuately it is only seen to advantage over the railway bridge now.

This, if I may be forgiven, is a 'mock' smock-mill, being a 'reconstruction' of the original which was burned down in the early 'thirties.

The white smock-tower (Plate 63), stands on the original brick base, but is much shorter than it should be. The second illustration shows the mill before the fire, and it appears to have been at least 8ft. higher, with a full set of spring sweeps (Plate 62). Nowadays this can only boast a set of stocks and whips plus a fan, with a missing blade, on the slender stage.

Whilst what we see today is not very closely related to the original, it is well worth noting the salient points of history and construction of the working predecessor.

Originally known as Barry's mill, after the early miller, one Frederick Barry, who was shown in Pigot's 1828/9 *Directory,* she is said to date from 1824.[1] The carved inscription F.B. 1824 was to be found within the smock at one time. In 1840 Mr. Barry was shown at his mill on Gibbet Marsh for tithes[2] and in 1844 the *Sussex Weekly Advertiser* carried the following notice:

> 4th March 1844
>
> To be sold by auction by Mr. Fuller:—A capital fantail windmill situated at Rye in Sussex, erected a few years since of the best materials, in the most substantial manner with five floors, two pairs of French stones and one pair of peak ditto, flour and dressing machines, iron machines and gear, smut machines and patent sweeps, constructed with all the modern improvements at the cost of £2,000. This mill is situated on the wharf near the 'Strand & Rye ferry' and a few yards from the intended railway terminus.

The patent sweeps were later changed to springs. Mr. John Smith and his family followed in 1848 and were millere here[3] up till 1879, when the mill was put up for sale again,[3] followed by George Standen from 1882 to 1899,[4] when a 9 h.p. vertical steam engine was also employed. By 1917, Mr. Webb,[5] a miller and merchant baker, had taken on the business. He had come to Rye when Moon and Co. finished.

In 1930, 'Gibbet Mill' as it was latterly known, was part of a working bakery leased by Mr. Webb. The upkeep was his responsibility, and after the fire, it was decided that the mill should be rebuilt. The outcome was that a shorter body devoid of floors and machinery was erected, the whole topped with a rather odd cap. This has an exaggerated profile, unlike the working original.

The 'New Mill' was erected in 1932 under the general supervision of Neves, the Heathfield millwright.[6] They made and fixed the sweeps to the wooden smock which had been erected by a firm of builders. The total cost to Mr. Webb was £800.[5] His grandson, Mr. Alan Webb, ran a bakery business on the premises until 1976, and with aid from the East Sussex County Council, in 1968 employed a firm of resin-renovation experts to coat the weather-boarding in white fibreglass. This gives this landmark a startling appearance on a stormy day (Plate 63). The premises are to be used as a pottery.

While here, the town of Rye, with its picturesque fishing harbour not far from this scene, is well worth exploring. The old cobbled streets and timber-framed buildings offer a genuine diversion into the past. The town, unlike the mill, is on a hill, which was surrounded by the sea. To the south-east there is the gentle meeting of the Tillingham and Eastern Rothers which flow as one to the sea, three miles away, across the Levels.

On the edge of one of the plain-like marshes of Sussex, providing a contrast to the rolling downland, the mill can be found on the north side of the railway bridge which crosses the river here. **[TQ 917-203]**

SHIPLEY (King's Mill) (Vincent's Mill)

White Smock-mill, with Fantail, maintained in Working Condition throughout its Life

Here is a beautifully-cared-for example of the millwright's craft (Plate 64) standing in an equally beautiful setting. The village of Shipley has every right to be proud of its windmill. Known widely as King's or Vincent's mill, it stands idyllically not far from the village church and only the length of the mill-field from a head-water of the River Adur. Here, the as yet small river wends its way slowly through the West Sussex Weald.

We are only two miles south of the A272, which carries holiday traffic on to Billingshurst and the West. Yet here, if we pause at the river's edge and take in the scene, we are out of the modern age. The mill, Mrs. Shipley, stands before us on a gentle rise. To our right, water meadows and grazing land accompany the river round its sweep to the south, below the church. If we return and look above the door in the north face of the base, we see the reason for her present good order. A stone plaque commemorates the presence here of Hilaire Belloc.

A famous Sussex poet and writer, Hilaire Belloc was first a countryman. He came here to live at Shipley in the house called 'Kingsland' (Terra-Regis in his book, *The Four Men*), behind the mill, in 1906. This was working up till 1926 (although latterly by steam only, I believe),[1] and Hilaire Belloc saw to it that she was, as far as possible, well maintained. After his death in 1953, a fund was set up in his name, to completely renovate and restore, the work being carried out by E. Hole and Son, over a period of some years at a cost of £4,500. Mr. Peter Powell, the son of the last miller, was able to grind wheat here during 1960.

Vincent's mill, one of the youngest in Sussex, was built in 1879 by Horsham millwrights called appropriately, Grist and Steele, for £2,500. Some of the machinery may be considerably older than the mill itself, for it is believed that components from an old mill near Coldwaltham were incorporated here. Mr. Friend Marten, the original owner, was extremely distressed at the high cost which had been estimated at £800.[3] He died in 1884, leaving his wife to continue with Mr. R. Wood as her miller until 1895.[1] Mrs. Marten then sold to Richard Vincent who was shown in 1898.[1] Then came Hilaire Belloc, whose family still own it, although West Sussex County Council are responsible for maintenance. Mr. Ernest Powell was miller to Mr. Vincent and Hilaire Belloc and was here until 1926.[4]

One of the first points noticed about this Wealden mill is its stoutness. It is extremely roomy inside, with a well-designed layout and ease of accessibility.

The octagonal conical cap is large with a petticoat to protect the iron dead curb from weather, and now sheeted over. This has two good-sized doors, one leading out on to the fanstage. Here, the six-bladed fan keeps the sweeps into the wind continuously, with a gentle clacking of its winding gears. The other door opens above the windshaft to facilitate lubrication and maintenance of the striking gear. The windshaft is of iron and carries a 7ft. Ø wooden clasp-arm brake wheel. The bin floor is surprisingly roomy, the middle taken up with vertical joists and the octagonal upright shaft, which is of oak and 15ins. across the flats. This carries an iron wallower above and great spur wheel below in the framing between stone and the spout floors. This underdrift arrangement makes life very much easier for the miller and stone dresser, all the stone drives being out of the way beneath the miller's feet. A fourth nut here was layshaft driven from the auxiliary steam engine situated in the adjacent buildings. This nut could be lifted into mesh with the great spur wheel after the windshaft had been put out of gear for working thus. The stone floor has three pairs of 4ft. Ø stones, two of French burrs and one of peaks (Plate 29), also a dresser. The spout floor has one set of governors controlling three sets of tentering gear above, also a smutter (Plate 33), an oat crusher, and, of course, the meal spouts from the stones, and the meal bins themselves. Spouts from the dresser hang nearby for flour sacks to be offered up. Here still on the spout floor which is 26ft. across at the corners, are two doors on opposite sides which lead out on to the gallery. This is the best place to appreciate the power developed by what must be the largest and perhaps the most extravagant mill in Sussex. As the 34ft. patent sweeps cleave past, they creak with the changing loads imposed on them as they pass the mill body and climb into the sky. As the wind rises, the weight may be lowered on the weight chain to allow the shutters to open a little, and so keep the mill running speed at the right level. These sweeps which are carefully balanced, using lead weights, are double shuttered, although the leading edge shutters are only half the length of those in the trailing edge, curiously one sweep has 13 bays and the other three have twelve. Control of these is through a notably executed set of sail rods and triangles connected to the central spider coupling (Plate 6). These, and the striking rod, lend an occasional clank to the running noises of the mill. The stocks are bolstered to the canister by a combination of straps and long wooden clamps.

The white octagonal wooden smock tower has a good supply of windows. Below the spout floor is the two-storey octagonal base, some 24ft. across corners inside the ground floor. This, tarred black, was built in brick by one John Summersell, who was shown in *Kelly* as a local builder. The engine house and granary are adjacent to the eastern side.

Above all other mills in Sussex, with the exception, perhaps, of Nutley post-mill, this one should be visited when open, and if possible when there is enough wind to turn the sweeps and perhaps grind. For here is the romance of Sussex milling in its setting as well as its mechanical sophistication if perhaps too young to remember the less leisurely (Plate 52) and older ways of other Sussex windmills.

Very out-of-the-way, and not visible until almost on the spot. The best way to approach this mill is to take the Cowfold to Billingshurst road A272, and continue westwards until we reach the little turning to the south which takes us to Shipley. The mill stands in the village itself to the south of a lane passing through.

[TQ 143-218]

WEST BLATCHINGTON

Preserved Black Smock-mill with Fantail, part of Farm Buildings

Here, high on the gently sloping southern scarp of the Downs, overlooking the English Channel, stands this very unusual and interesting little smock-mill. We can easily miss this site, for today it is completely surrounded by houses. Hove Corporation acquired this from the Marquis of Abergavenny for £3,400 in 1937,[1] and are responsible for the upkeep of the mill, permanently retired as it is.

This, in its working day, was a utility mill (Plate 66), part of Blatchington farm. As can be seen, this stands on a three-storey square base, built of that most traditional of Sussex building materials, chalk flints, second only perhaps to Horsham stone. The base, which has a slight batter, appears to have been raised through the large barn already in existence, which abutted north and south, incorporating one of the existing structure walls. A flanking barn on the west, seems to have been added a little later. All that remains of these and a granary building to the west is the western abutment (Plate 65). The north barn was demolished and the south burned down in June 1936,[2] the whole was in a farmyard proper as opposed to the usual mill field or yard.

This mill has often been mistakenly dated around 1720 and associated with smuggling which was rife in the mid-18th century. However, whilst some of the timbers in the cap look very old, it appears that she is 100 years younger,[3] the first evidence of this nostalgically rural complex is its appearance on Greenwood's map of Sussex dated 1823, and a contemporary water-colour by John Constable painted during his stay at Brighton, and dated 5 November 1825.[4] Strangely, a map of this part of the Abergavenny Estates, dated 1829,[5] and a subsequent tithe map, do not show the windmill, although the farm buildings on which it stood are clearly shown. This fine seamark was used for grinding flour and feed, for which it was equipped with two pairs of 4ft. underdrift stones and as a source of power for farm machinery for the Hodson[8] family who were well known as millers and farmers, and they were shown until John Brown came in 1887.[8] A Mr. Whittington was the last to work the mill in 1897.[6]

She was originally equipped with common sweeps as Constable's painting shows. These were later changed to patents. It was when two of these were badly damaged that the mill finished work. They were removed several years later and the mill stood with two sweeps and four stocks until the first restoration was started in 1938/9 by Neves[6] of Heathfield.

The slender smock tower is empty today apart from flooring and the upright shaft. The stones were removed in 1937. These were situated in the base of the smock tower at the level of the outside stage. Below this floor, the composite great spur wheel (Plate 21), which is supported by an impressive 'A' frame hursting, drove the two stone nuts and a third pinion which, via a vertical shaft, bevel gear and pulleys powered the layshaft in the barns either side of the base. Here a threshing-machine and chaff-cutter were once situated. This mill may well have replaced a horse or ox gin previously used.

Today we find the spokes of the six-bladed fan are carried by a well-restored fanstage. The hooded cap, almost rounded over, is somewhat narrower than the curb which is perfectly housed under a very tidy petticoat (here again Constable appears to record a beehive cap in 1825). Inside, the iron wind-shaft still carries the iron brakewheel, though minus cogs. This has several steel tie-rods passing between the spokes which prevent it from turning (although it is held by a screw-down brake fitted by Holloway). These were placed here to pull the sunken breast beam back into the mill some time ago. Below the cap frame, whose timbers are so aged in appearance as to belie the mill's youth, the wallower remains atop the 4½in.-square iron upright shaft.

Apart from the more obvious fact that the small smock of this interesting mill is only six-sided, it is notable in that some sensible and effective iron work was incorporated. Below the iron curb, a structure of cast diagonal brackets is worked into the top of the smock, preventing the curb from distorting badly and sinking as is so often the case. Likewise, each of the floors is supported by cast iron radial socket work, which forms a steady bearing for the upright shaft and replaced the large binding timbers usually fitted. From the stone floor, the stage is gained by a door in the north side of the smock, this allowing the miller to service the sweeps and set the striking gear. A steady rope is provided around the smock which I found to be most reassuring when standing here on a windy day. One Rounderby,[7] who tarred mills and farm buildings in the area, no doubt shared my experience on occasions, for he would conduct his domestic quarrels in brave tones from this vantage point. His wife being timid of heights could, nevertheless, console herself with the fact that he must at some time come down.

Another notable feature, shared with the bygone pumping mill at East Blatchington, near Seaford, is that the sweeps are carried on a substantial cross-iron, nicely webbed back to the windshaft. The stocks are bolted back to this and are all in the same plane, held with bolts and clamp irons. The millwright, his name is unfortunately lost to us as with the majority of mills, must have been pleased with his work which even today is seen as well built and in the best traditions of later milling practise. It would seem that, from Constable's painting, the mill was re-worked at some time during its life.

Recent enthusiastic interest in West Blatchington mill, which is now a Class II-listed building, has prompted an organised effort by Mr. Wood of Hove Borough Planning department to revive the structure in some way. Sadly, she cannot work at present, but it has been suggested that this, and the buildings, might make a fine museum.

Here I must make an aside for all lovers of Sussex views. If we approach the site by taking the road from Saddlescombe which crosses the top of the Downs, continuing until the Municipal golf course comes into view on the left, we see to the south-east Patcham tower-mill on the rise up from Waterhall, and a mile to the west of this, with the sea as background, the black sweeps of our present quarry, West Blatchington. Again, behind and to the north-east are the two mills at Clayton, some four miles away. We must not, of course, neglect the magnificence of the High Downland all about us. The sight of four mills, all of which are still standing, although not at work, is pleasing enough to any windmillian.

From Dyke Road crossroads, we take the steep downward path of King George VI Avenue towards Hove, and on rising, we gain the top of the next rise. Here, we can see the mill to the left down a short side road.

It is difficult to imagine that this areas as it was 80 years ago, for this mill was the centre of attraction in a bygone rural scene.

[TQ 279-068]

WEST CHILTINGTON (Meeten's Mill)

Converted Black Smock-mill with Fantail

This sturdily-built black smock-mill (Plate 67), not far to the north-west of Washington was, according to a little booklet on the village,[1] 'moved here about 1830 from a site near Monkmead'.

The first positive evidence of this mill's existence is her appearnce on the 1840 tithe map[2] of West Chiltington, the schedule of which shows W. Hammond. Later, in 1844, a notice appeared in the *Sussex Advertiser* which read: 'To be sold; excellent built smock mill at West Chiltington, now in the occupation of Mr W Hammond @ £4- / annum the mill being modern.'[3]

It has been suggested that this mill was erected in 1688, evidence of which was said to be found on a date stone. The only such stone to survive appears to read A P 88 or 38, the latter taken with traditional local knowledge[4] and existing map evidence makes a date of 1838 much more acceptable.

The Willmer family worked the mill from 1845[5] and in 1865 Coopers of Henfield were employed to refit the mill. This was discovered when the mill was cleared of machinery in 1922. Mr. Powell of Coolham water-mill bought the stones and remembered seeing the millwright's mark and date on various pieces of the iron machinery.[6]

The sandstone tarred base is a fine one and has three floors. It is octagonal and above the first floor, has a stage for servicing the sweeps and winding the mill by hand. Brick-faced piers were built on to the base to add strength and support the stage.

A hand-winding wheel once used has gone and was at some time replaced by a stage-mounted fan. This was a pretty white one, now unfortunately amiss. The large hooded cap is now fixed.

Of the sweeps, two remain. These were originally shuttered, and controlled by springs which are still in place. The other two sweeps were cut off near the canister some years ago. Maintenance of these is rather tricky, for they cannot turn full circle because, at the front of the mill, there is an extension. This is part of its conversion into living accommodation some 56 years ago. It is usually necessary to turn them to the upright position for removal of the stocks and sweeps for repair work.

The miller here from 1898-1918[5] was Mr. Meeten, who saw the mill finally stopped in 1922. Villagers, naturally enough, knew this as Meeten's mill, and when working, the mill drove three pairs of stones and other machinery

(Plate 68), all now gone. An auxiliary engine was also used latterly. The Meeten family was at one time connected with local smuggling, and no doubt a local mill here played its part in earlier times. Nowadays, West Chiltington mill is in stately retirement. She stands on the west of the village proper in a sunken lane and overlooking one of the best-kept villages in the county. Situated four miles from the foot of the Downs in this part of West Sussex with its ancient hamlets all around us, the village of West Chiltington and its mill is three miles east of Pulborough on a minor road. The tower remains of Nutbourne mentioned below is easily reached from here. [TQ 085-181]

This is the final note on Sussex smock-mills here, although in the later sections, I have included the remains which are still recognisable. Nearly half of the remaining smock-mills are converted into accommodation and this has, of course, preserved the outer fabric at least where demolition or decay would most certainly have taken over.

I must add here that the smock-mill known as 'Shiremark' is often mentioned in previous writings on Sussex windmills, and indeed by Hilaire Belloc in *The Four Men.* Here, she is claimed as the guardian of the gate to Paradise (Sussex), where the angel with the flaming sword will stand on the day of judgement.

Known variously as 'Shiremark', Kingsfold or Capel mill, it did, in fact, stand just over the border in Surrey. Today all that remains is the base and the windshaft. This stood derelict for many years in a dense thicket until, bereft of sweeps and good deal of weatherboarding, she was completely destroyed by fire in October 1972. An account of this mill, including sectional drawings, is th be found in *Windmills of Surrey and Inner London,* by Farries and Mason.

III

TOWER-MILLS PROPER

ARUNDEL (South Marsh Mill)

Externally Renovated Black Tower-mill having had a Fantail

Here, we are on the east bank of the River Arun on its way to Littlehampton. The town stands to the north, dominated by the famous castle which was the seat of the late Duke of Norfolk. All that remains of the mill is a sweep-less tower topped with an octagonal mock-up placed here in 1936 and quite unlike the original beehive cap (Plate 69).

The brick tower has five floors, these once being gear, bin, stone, spout, and ground, and worked three pairs of underdrift stones, one of peaks, and two of burrs.

South Marsh mill, as it is known, was erected in 1830, before the railways took hold in Sussex, and the lessees, Messrs. Dendy and Pellet,[1] would have seen many sailing barges passing up and down the river. Some of these would have sailed to Littlehampton with coal and other essentials, and then been drawn up-river by horses, as inland river barges were, to the busy quay here, and further north via the Wey and Arun canal in its heyday. The fearful current that occurs when the tide is out at the river's mouth must have caused many problems for the mariners of the day.

The business was worked by the Bartlett family from 1840, when Henry Bartlett and William Watkins were shown,[2] and 1922 (Plate 70). The mill stopped working by wind in 1915[3] when a fierce gale damaged the sweeps and windshaft beyond repair. The brick tower here has a basic fault in that the windows are one above the other and form a natural fracture line from top to bottom. This, perhaps, contributed to the deterioration by causing spreading, and would most certainly have distorted the curb upon which the cap rotated. The sheared iron windshaft remains behind the penthouse extension, but the other machinery was removed from the decaying remains in 1940/41[1] to further the war effort no doubt.

This, situated as it is on the flats of the Arun Valley, originally had a very good draw so to speak, most of the area to the sea being served by it in the heyday of the windmill before the roller power-mill came into its own around the turn of the century.

Whilst in the area, time should be taken to explore the Arun north of the town. The villages of North and South Stoke, Burpham and Amberley afford the most rewarding walks in beautiful countryside overlooked by Arundel Park on the west. Here, the Downs part to allow the combined waters of the western Rother and the Upper Arun to flow through.

Easily found on the east bank of the river at the end of a short road which leads southwards off the A27 just before crossing the new Arun bridge.

[TQ 013-063]

BARNHAM

Sweep-less Black Tower-mill having had a White Cap and fantail, now part of a Power-mill

Like Punnetts Town smock-mill, this was a pleasant surprise. Earlier writings on Sussex windmills indicated that the mill was disused, although almost in working trim. Some 40 years have passed since then, and today the mill still has most of its internal working machinery, although minus sweeps and fantail.

Approaching Barnham from the direction of Littlehampton, we pass the disused Royal Naval Air Station at Ford. The road crosses the runway which is now endowed with large heaps of farmyard manure and huge baled straw stacks. After passing through Yapton, we see the mill a mile or so north-west. I had expected that the tower would have gone the way of others and become an ornamental residence. However, this still forms part of the feeds business: John Baker, Barnham Ltd. This was established in 1880/2[2] as 'Maurice and John Baker'. Subsequently, John is recorded and the business remained in his name, although aided by his eldest son, Percy,[2] from 1910 until 1916, when the younger son, Len, took over. Len Baker continued until his death in 1945.[2] Mr. R. C. Reynolds took on the business in partnership with Len's three sisters. When I arrived here Mr. Reynolds, who is the nephew of Mr. John Baker, was only too pleased to explain the working of the mill, and allowed me to spend time going over the plant. Today oats are rolled in a small mill on the spout floor; barley, wheat, and other feedstuffs are reduced to meal for provender in two small hammer-mills situated in the granary buildings extended around the base of the windmill tower (Plate 74). The resulting meal is blown up into the spout floor through pipework, to be dressed and mixed in modern machinery. From here it is passed down to the roundel floor to be sacked up. Situated in the other buildings, which once housed a Tattersall's Midget Maxima Breakroll flour mill, are a roller-mill and pelleting machinery. Approximately 20 tons of various feedstuffs for cattle, pigs and chicken, etc., can be produced here in a week. The inside of the mill has a liberal coating of meal as do the men who work here (Plate 22). Each type of feed has its own nose so to speak, some sweeter than others (Plate 55f). The pelleted pig-nuts have the particularly pungent odour of cod liver oil.

Mr. John Baker asserted that this mill was erected in 1790 to replace Feaver's post-mill. (This stood about a quarter of a mile west of here.) However, there is a contemporary news item which records the fall of an old post-mill at Barnham in October 1827,[3] and another which is dated 23 January 1830[4] posting the sale of a mill here: 'This substantially built cylindrical mill recently built without regard to expense and working 2 pairs of stones'. The Feaver family did in fact work the tower mill until 1880 or so.

Barnham mill originally had two spring sweeps and two commons, and worked thus until 1890 when J. W. Holloway of Shoreham was employed to modernise the whole business for Mr. Baker. The millwright had the tower heightened by 18ins., then fitted a new cap and larger patent sweeps. All the machinery appears to date from that time. The brakewheel here has a family resemblance to the one fitted by Holloways at Medmerry mill, Selsey, when they carried out similar work there.

The tarred four-storey tower is constructed of flint and stone cemented together. The outside is well rendered and decorated with ornamental mortar lines and a large advertisement panel. Today the sweeps and fanstage that the cap supported are gone, removed in August 1958.[5] Mr. Reynolds remembers the mill working by wind in the early '20s, after which time the original power-mill took over entirely (Plate 75). A 49 h.p. gas engine drove two pairs of emery-faced stones on the stone floor from below via a layshaft and an anxiliary gear wheel. Today the stones remain as they were left some years ago when the great spur wheel was damaged beyond reasonable repair. A third pair fitted by Holloways were removed. In the cap we find the large iron windshaft which is unusual in that the canister is flanged and bolted to the shaft, having been cast separately. The large iron brake wheel at first appears to be one with the rest of the shaft (Plate 21c). Close inspection reveals that this was cast in halves, clamped to a round shoulder and made secure with a stout key. The brake here is applied by Holloway's geared screw as opposed to the more usual lever mechanism. This arrangement is also to be found at Patcham tower-mill, which was built by Holloway of Shoreham. The upright shaft is iron and also in two separate pieces supported by four bearings. The wallower, which has been removed, drove the lower half via a large 'oldham' type universal coupling. In the stone floor we find that the bins fed two pairs of underdrift emery-faced stones. There was also a dresser fitted here I believe. Below again on the spout floor, the upright shaft terminates in a great spur wheel and a lower bevel wheel for auxiliary machinery (Plate 22). The stone nuts are held out of mesh today by jack-rings as is usual in a later underdrift arrangement (Plate 28). The stone tentering seen here today is very basic, being applied via a single lever with a fulcrum pin at one end and the tenter screw at the other. Perhaps the governor and steel yard arrangement were found to be too 'fussy' for the type of grist work done by the modern power-driven stones.

Originally the mill boasted a gallery around the outside of the tower at spout floor level. From this the sweeps were set through the usual weight chain and wheel. Unfortunately, the supporting timbers for the staging were damaged during the last war when the army used the mill as an observation post, an obvious choice, for this mill commands an unhindered outlook over many miles of the coastal plain and out to sea.

One wonders how long it will be before this survivor of milling in Sussex falls to the march of the big feed companies, as others have. This must be the last of those associated with windmilling left to us in this county today.

I add that in 1978 Mr. Reynolds gave up milling on the premises and now buys his feedstuff from Ranks, still mixing as required.

Easily found to the south of the A2024 Littlehampton to Chichester road in the village of Barnham itself. **[SU 968–039]**

JACK–CLAYTON

Partially Restored Black Tower-mill with White Cap and Fantail

Jack is understood to have been built in 1866[1] to replace the original Duncton post-mill that stood on this attractive part of the Sussex Downs (Plate 72). Jack's tower was built against the post-mill roundhouse, and a doorway provided between the two. It has been suggested that Dunction mill was still working whilst Jack was being erected. A cursory glance at Plate 71 will show the reader that this plainly quite absurd.

The miller of the day, Mr. J. Mitchell,[2] had this large tower-mill built to a height of five floors. These were from the top down: gear floor, 13ft. inside diameter; bin floor; stone floor; spout floor; and ground floor, 22ft. inside diameter. The spout floor, where the final result of grinding and dressing is sorted into sacks, has a railed staging around it with fine ornamental cast iron supports (Plate 71). Two doors open on to this, facing north and south. The combined height of these floors puts the curb of the tower at something over 44ft. above the ground. Here we see that the southern side was metal clad to keep damp at bay. This was carried out in 1934 after numerous earlier alternatives were found wanting. It is thought that Cooper of Henfield,[3] known for his fine iron work, was Jack's millwright and that Holloways of Shoreham, and Whitington of Brighton, were involved here in later years.

Three pairs of stones appear to have been underdrift with an auxiliary power take off in the spout floor. Evidence in the flooring indicates this, but nothing remains today, the tower having been gutted of machinery after the mill's retirement. Here the three were fitted apparently with room for a fourth should they be required later. The stones were 4ft. in diameter, one pair of peaks and two of French burrs. There are three pairs of stones that were at one time or another used in this mill laid out in a path after Jack ceased work.

The brothers Joseph and Charles[9] Hammond took on the mills in 1871-2. Charles had Jack fitted with up-to-date auxiliary machinery, including an oat roller in the old Duncton roundhouse. A tunnel was dug between Jack and the granary buildings to provide a dry walkway. He is shown at Clayton in *Kelly's Directories* until his death in 1903, when a Mr. A. Wood took on the lease with Mr. Guy as his miller. They continued for three or four years before work finished for good. After this, the archaeologist, E. A. Martin, used the base as a holiday home in 1908, 1909 and 1910.

Charles Hammond was always eager to improve his business, and, with this in mind, he patented a novel method of sweep control in 1873.[3] This was in the form of a large centrifugal governor combined with Cubitt's weight and chain gear. The mechanism was fitted into the cap of Jack (as the patent shows) and at Hurstmonceux post-mill, where it may be seen today (see page 46). Joseph Hammond was working at Hurstmonceux between 1878 and 1892, and fitted this to help promote the idea, I imagine (Fig IV).

Jack's original sweeps were a fine set of double-shuttered patents. As their speed varied from a pre-determined optimum, the governor would have automatically opened the shutters through the striking gear. Relying on sweep speed rather than actual wind pressure is perhaps the ultimate in mill control, for the speed at which the various pieces of machinery are driven by the sweeps is quite critical and must be carefully watched.

The aim of Mr. Hammond's governor would have been to make the mill as automatic as possible. Nature has its own ideas on these matters. The wind cannot be tamed and the miller would, in fact, have to adjust his work-load to suit the wind rather than the reverse, especially after perhaps a week of so of near still air.

The white beehive cap is large, some 16ft. diameter inside, topped by a ball finial. It is extremely weathertight, being double shingled in what looks like ash lapping. From the stage here, gained through a neat storm door in the back of the cap, there is an impressive view over Wolstonbury Hill to the west and the Weald beyond Clayton village 400ft. below us. The miller would have to be steady and have a good head for heights, for he would come out on to this stage to check and oil the fan bearings and gears which drove the cap into the wind. Alas! now the five-blade fan has gone, damaged, as were two of the sweeps in gales in 1909.[5] At this time, Mr. Martin was in residence and tells us of his stay in his book *Life in a Sussex Windmill.* He actually lived in the first two floors and the adjoining roundhouse of the original Duncton post-mill. Romantic though it may seem, he gives a different picture of life in the mill. The furnishing was something of a problem, as can be imagined, the carpets and fittings of a 12ft.-square room are rather lost in one 20ft. or so in diameter with an iron pillar in the centre. The pictures had to be pinned back to the sloping walls, and Mr. Martin recalled with amusement how, when opening the door on a windy day, every loose object skated and whirled around the room. Not so amusing were the insects and mice which inhabited the mill at night, to the groan and clank of the cap as the fan kept the weatherbeaten sweeps into the wind. Passersby would call upon him, after peering through the windows, and insist that this was their right, and that he should serve teas. I am sure that the spirit of the Downs more than compensated for these troubles, for he seemed happy to be here and did, in fact, make several archaeological studies during his stay.

The two remaining sweeps were finally removed and the Anson family, who were to stay for over 40 years, came to live in 1911. They had the machinery removed and made the tower and granary buildings more habitable. Mrs. Anson's father, Captain Cooper Coles, had been a distinguished sailor, and when her husband, Captain Walter Anson, R.N., died, she had the stone floor fitted out as a chapel in memory of her seafaring menfolk (Plate 73).

In the early days, the two mills still stood on open Downland and later, Captain and Mrs. Anson cultivated the garden and hedges which had started to grow, converting the old granary building into a residence.[7]

Before his death in 1978, Mr. Henry Longhurst, the well-known writer and golfing correspondent, lived at the mills. He came in 1953 and made a modern home built to replace the old buildings in 1961.

In recent years, Jack has been completely re-painted and renovated. Also, in 1973, a film company approached Mr. Longhurst and the Mid-Sussex District Council, who care for the two mills. They wished to use the mills as setting for a film, *The Black Windmill*, and agreed to pay £3,000[8] to have a beautifully-made set of basic frames put up with hew stocks. This work was carried out in July 1973 by Mr. Peter Stenning, of E. Hole and Son, so that Jack was able, after some 64 years or so, to boast a set of sweeps, giving him a very workmanlike masculine appearance once more (Plate 72).

As for Jill post-mill, we gain this famous spot by the mill lane from Clayton Hill. [TQ 304-134]

HALNAKER (Boxgrove)

Externally Renovated Tower-mill with Hand Wheel

This attractive old tower-mill stands on Halnaker Hill, an upthrust of Down, 420ft. high and four miles north-east of Chichester (Plate 76). The site must be visited on a clear sunny day, for it overlooks the whole of the Selsey peninsula to the south and over this to the west, the Isle of Wight is easily seen 20 miles away. Behind us are the forests of Goodwood Park, West Dean, and Upwaltham. These are mostly Forestry Commission plantations, the majority composed of concentrated conifers with beeches intermingled. The conifers are grown for pulp wood and felled to keep up with the ever-growing demand for paper and hardboard, leaving the beeches to continue growing until they are felled in old age, which can often be 200 years.

Returning to our present subject, the basic sweep frames of the mill here face Chichester now, set in this position during 1934 when the cap and tower were restored by Neves of Heathfield as a memorial to the wife of Sir William Bird. Inside the brick tower, a plaque giving the names of the workmen involved, also tells us that Mr. Rex Wailes acted as consulting engineer. He also advised in this capacity at Chailey when the smock-mill was restored in 1936.

Halnaker mill was built for the Duke of Richmond as the feudal mill of the Goodwood estates. Here the tenant farmers and alike were able to have their harvest or gleanings converted into wheatmeal for their own use. The wind from all quarters is unhindered here, and the sweeps must have been able to turn when many mills were idle.

The first note that I have of a mill at Halnaker is from the Goodwood Estate records. Detail from a Henry VIII Crown lease of 1540,[1] shows tithe and tenancy details for the manor of 'Halfnaked' in which: 'the farmer of the windmill of said manor pays yearly for rent—16 quarters of corn—with the multure of all the grain sent to the mill by farmers and millers of the same mill to be ground'. (Multure is Old English for the toll taken in kind for grinding.) Further references show that a succession of windmills stood here from that time up till the present day. As far as is known, every map of Sussex that showed windmills included one at Halnaker. The date of the mill known to us now has proved illusive. Generally put at *c.* 1750, Halmaker mill or 'Halnecker' as she was named on earlier maps, is first positively shown as a tower-mill on Goodwood Estate survey map prepared for the Duke of Richmond, *c.* 1780.[2] It has been suggested that this mill dates from Queen Anne's reign, 1702-1714, and whilst it is very probable that there was one on the hill at this time, there is nothing to suggest that it is the one that we see today. The brick-built tower-mill did not gain prominence in any case before the latter half of the 18th century in England. It is certain that Halnaker tower-mill qualifies as the oldest tower-mill in Sussex.

A reference to a tenant miller is given in the estate conveyance of 1810[3] and shows John Hervey with mill and 17 acres of land. Charles Adams was shown here from 1839[4] till 1870.[7] Mr. G. R. Watkins was the last miller here.[4] His family worked the mill from 1868, when a mill-house still stood close by, until 1905 when lightning struck her, splitting the wooden windshaft and damaging the cap. The owners were not inclined to have this put right, and so the mill became derelict.

When working, she had two pairs of overdrift stones, powered latterly by common sweeps via a wooden windshaft, and had four floors. It is obvious to anyone standing here that the fifth floor often included in smock- and tower-mills to give increased height was not needed, the command of the wind is perfect.

Originally hand-winded, using a handwheel and chain,[5] the mill finally had a fantail of which there is now no trace. The handwheel may still be seen enclosed in a rather fancy stage at the back of the cap.

When Halnaker was renovated, Neves had a lot of work to do, for by this time, the mill was quite derelict (Plate 77). The new common sweep frames, which incidentally were renewed once more in 1956 by E. Hole and Son, were fitted to the iron windshaft and brake wheel taken from Punnett's town saw smock-mill, mentioned earlier.[6] This windshaft, which was in two pieces bolted together, proved too short and so Neves added a spacing piece between the flanged faces to bring it right. A 16-sided beehive cap covers the now empty brick tower. This has had its windows filled in, and of the outer covering of red tiles about 50 per cent. remain.

Those interested in Romano-Sussex will find in the area here enough to satisfy, for to begin with, the Roman military road known as Stane Street, runs across Halnaker Hill on its straight course from Chichester or Noviomagus Regnensium as it was then called. Here it passes over the Downs in a north-easterly direction to Pulborough, where its alignment changes, taking it on through Adversane and Billingshurst into Surrey on its path to London, surviving today as part of the A29.

At Fishbourne there are the recently-excavated remains of the Roman Palace, and close to Stane Street itself in a perfect setting, are the remains of the Roman Villa at Bignor. These have very fine examples of mosaic flooring, also archaeological evidence of the occupation of this county.

The best and most rewarding way of approaching this attractive mill is to take the A285 out of Chichester, which, incidentally, is also part of Stane Street, and pass through Halnaker village. A little further on, the road skirts the bottom of the hill till a lay-by is reached. From here there is a public footpath which takes us up to the mill from the south-east. The half-mile walk ends with the magnificent views described above, and, of course, the windmill to keep us company. [SU 920-097]

NUTBOURNE

Derelict Black Stone Tower-mill having had a Fantail.

We return to the quiet West Sussex Weald, some 12 miles north-east of our last quarry, still within two miles of Stane Street. As may be seen in the photograph, this is still a mill proper, but strictly speaking, only just. However, it provides a good home for a flock of pigeons (Plate 78).

There were three mills here at Nutbourne in the late 1890s,[1] this tower-mill which was built in the 1850s,[2] and two water-mills. Only one, a water-mill, is shown on the first edition of the Ordnance Survey map. A later map dated 1861,[3] shows the windmill as well. In 1890, Messrs. Reed and Stillwell ran the mills, as had a series of farmer millers, and a bakery business[4] which by 1900 was much reduced, no doubt pressed hard by the changing fortunes of natural power-milling against large power-mills. The mill is said to have been used in 1908 as a summer residence by Dr. Barnado's Homes,[1] but it seems unlikely, as most of the machinery remained within the mill until recent years. Perhaps this refers to the ground floor only.

The tower is, like East Wittering, a tribute to master masonry. It is beautifully executed in sandstone which, incidentally, was heavily tarred, traces of which remain as does the whitewash inside. The windows and doors are finished in larger stone blocks, well keyed into the surrounding stonework. The sockets for the staging are seen just below the spout floor doorway. Similarly, here as at East Wittering mentioned later, the wooden curb and rack is still in place, its cogs gradually rotting and falling out. Then balanced on the top, a criss-cross of aged-displaced timbers. At this time, the iron windshaft hangs from the top of the tower (Plate 78). The tail beam having collapsed, the heavy iron windshaft has fallen tail-end down, checked only by the stocks, which now form a horizontal cross, giving the outward appearance of a giant windlass. There must be about three-quarters of a ton weight on each arm now. Judging by the fact that the mill has not been used for at least 84 years, it seems likely that these are original and approximately 125 years old, which speaks well of the timber used. Set against this, the fact that a stock broke clean off at Polegate, having been up for only seven years.

Remains of the cap sheers are still in evidence, as are the two main fantail supports. The fan itself blew off just before the windmill stopped work, after a seemingly short working life in 1893/4. The cap proper was still recognisable as a Sussex beehive pattern some 40 years ago, and has fallen to pieces a little at a time ever since (Plate 79).

Inside the mill, the remains of the five floors and some of the machinery, most of which has crashed to the ground, could still be seen recently. This worked two pairs of underdrift stones which were taken out many years ago, driven, I believe, by patent sweeps.

To find this derelict, we must walk about a quarter of a mile down a picturesque field lane which passes over a mill-pond embankment. The remains of the water-mill itself can be seen to the right of this. A little further on, with orchards on both sides, the tower stands literally in the hedge of a field. The surrounding terrain gave one the impression that wind was not plentiful here, unlike the power for the water-mill near at hand. This may have hastened this sad old gentleman's early demise. **[TQ 078–189]**

NYETIMBER (Pagham)

Ivy-covered Derelict Tower-mill having had a Fantail

There is nothing about this mill to recommend it. Totally inaccessible, the brick tower remains, partially burned out by fire in 1962, although the windshaft can still be seen, its canister almost lobbed over the curb.

The whole is enclosed in a jungle-like thicket to the west of the road going south into Nyetimber itself. The almost camouflaged top of the tower may be seen if we take a footpath which skirts the southern edge of this thicket. This brings us to a caravan site and immediately we feel that we should have taken another path, but not so, for our objective rises above the eastern hedge wall here (Plate 80).

Not an easy one to find now, this West Sussex tower-mill is understood to have been built in the early 1840s for William Adams.[2] When working, the mill had patent sweeps which drove two pairs of French burrs and a flour dresser. Apparently no work for provender was carried out here. The upright shaft drove a set of governors which controlled the rate of feed to the dresser, an unusual feature.[3] The mill itself could be driven by an auxiliary engine and had a large take-off pulley for belt drive. Latterly, this was powered by a 16 h.p. Robey and Co. stationary steam engine.[4] This, I imagine, would be enough for one pair of stones and dressing tackle only. Nyetimeber, like Navarino South mill at Worthing, had an ogee cap which is the Sussex beehive pattern with a reverse-curved pinnacle, similar to a flat-bottomed onion. This carried a six-bladed fan and was sheathed with copper at one time. The cement-faced tower, which is now smothered in foliage, was tarred against the prevailing weather (Plate 81).

Mr. William Adams had the mill from the early 1840s until 1905.[5] Subsequently worked by a succession of millers, the last of whom was Mr. W. Prior, it finished for good in 1915 when the machinery was damaged after becoming tail winded. The sweeps survived until 1927[6] when lightning struck the mill and damaged these. Mr. Durman, whose family leased the mill for many years, used the ground floor as a carpenter's shop at one time. However, since the mid-'thirties, this remain has been neglected and decay taken its unhindered course.

Here on the Selsey peninsula, we find other things to interest us, including the three mills within easy reach during a day here. These are Selsey, and East Wittering tower-mills, and Earnley smock-mill. [SZ 892–988]

PATCHAM (Waterhall Mill)

Converted Cement-faced Tower-mill with Fantail

A Downland landmark for miles around, this very picturesque mill stands on Mill Hill behind West Dean. The road that passes here was a narrow lane some 90 years ago. Today it is a busy thoroughfare, taking traffic from the A23 at Patcham across to the A27 and the new Adur Bridge via Southwick. A very good water-colour of the mill in its early days is to be found in *Windmills in Sussex,* by A. Foord-Hughes.

Mill Road rises from below the main London to Brighton railway line to a point some 450ft. high at Dyke Road crossroads, where we have an extensive view of the English Channel, the town of Brighton, and over to Shoreham, one of the main ports of Sussex. Colliers, timber boats, oil tankers, and sand dredgers constantly pass in to the Adur estuary and on through the locks into calm water.

The scene from the crossroads is a marked contrast to that mentioned under bygone West Blatchington, with its almost entirely rural viewpoint of this locale. Turning eastwards, we face the mill, its white tower and sweeps contrasting against the rolling Downland background to the east (Plate 84).

The mill stands on the way down the hill, 100ft. below the crossroads. Built in 1884/5[1] for Mr. Joseph Harris. Here was an enterprising man—he decided, in his capacity as a local baker, to supply the area with home-baked bread made from locally grown and ground wheat. In fact he worked very hard in his mill, to the extent that one night in 1903,[4] when the wind was blowing hard, he got up and went to the mill on horseback, to make the most of the weather. The sad result was that he caught pneumonia and died a little later. His wife and family continued until 1924. Mr. Bull was their miller in later years.[2]

This was the last windmill to be built in Sussex, not, however, the last in England, as is sometimes suggested.

The Shoreham firm of millwrights, J. W. Holloway and Son, were employed to fit out the cement-faced tower which had been erected by Hubbard of Preston. The work occupied almost exactly one year, Mr. Hubbard commencing in February 1884 and Mr. Holloway finishing in the same month of 1885, at a total cost of £1,400. The fitting out with up-to-date iron machinery claimed £1,100 of this.

Well preserved and tastefully converted into a modern home, Patcham or Waterhall mill as it has been locally known, is of medium size. The tower stands 40ft. high and is 19ft. ø inside at the base, tapering to 13ft. ø inside at the curb containing gear, bin, stone, spout and ground floors, and is capped with an unusually roomy pepper-pot dome. Holloway fitted a Sussex pattern fanstage which carries a five-blade fan. Supported by the sheers of the stage are the weight wheel and a smaller wheel from which a brake chain was suspended. This operated Holloway's screw-down brake through bevel gears and shafting. Inside the cap we can see (Plate 19) that this rotated a bevel-geared anchor nut on the stud end of the brake band, giving a high clamping force without being too abrupt. A conventinal lever brake could, if misapplied, weaken or even break an ageing stock. The screw-down mechanism which, if not unique to Sussex, is very unusual in England. Other examples fitted by Holloway are to be seen at Barnham tower-mill and West Blatchington smock-mill. The brakewheel itself is of iron, split and bolted and clamped to a box centre which is in turn clamped and keyed to a round locating shoulder on the iron windshaft. Between the neck and shoulder and back to the tail, the shaft is nicely ribbed. A small stage was provided here, below that of the fan (Plate 15) for servicing the running gear. To prevent the weight chain thrashing about in a high wind, a steady-pole was also fitted. This was necessary as the traditional staging at spout floor level was not provided and the chains therefore hung almost to the ground. A black-painted iron strengthening hoop was fitted around the tower when cracks began to appear sometime after the mill's construction.

The original sweeps, fully shuttered with no leading boards, had a rather narrow full row of shutters in their place. These were controlled by a variation of the striking linkage attributed to Cheale of Lewes. If we look at what now remains, we see that instead of the solid spider iron or cross that is used in Cubitt's gear, a spring cross was fitted (Plate 85), thus giving each sweep a little independence of shutter movement, whilst retaining the overall influence imparted by the 'patent' system. Cheale's patent proper actually employed a full set of springs between the spider and the shutter bars of Cubitt's gear (Plate 8). Traditional millers were always sorry to see the shutters held open as patent sweeps came down out of the wind.

Inside the tower, three pairs of stones were underdriven. Priority was given to flour production, with two pairs of 4ft. French burrs and one pair of 4ft. 2ins. peaks. As may still be seen, a fourth nut was driven by the great spur wheel for an oat-crusher and a dresser on the floor below. The peak stones were used for grinding barley and pig corn. Some of the dressing tackle was brought here by Mr. Harris from the smock-mill at Chailey South when this ceased work in 1887 or so. Today the stones are boarded over under the floor of one of the bedrooms within the tower.

The Harris family gave up the mill in 1924, and in 1926 sold the property to Mr. C. Stubbs for £150.[3] The windmill itself was considered a liability at the time. However, the following note illustrates the changing value of this initially unattractive proposition. Mr. Stubbs proceeded to build the original cottage here, and over a period of some years incurred considerable expense. By 1935, the mill had become somewhat neglected and the whole for sale once more. The next to come, Mr. and Mrs. Stephen Easter, paid £1,900 for the mill and cottage. At that time, the tower was enclosed by a low hedge which today hides the lower half completely. After the war in 1946, she was for sale once again and was bought for £3,500 in the following year. The property, now an attractive windmill and cottage, appreciated to the extent that by 1961, £8,000 was the asking price. Miss Honoria Plesch purchased for a little less in 1963 and spent as much again converting the windmill into living accommodation without removing the machinery. She then took up residence in the windmill and in 1965 sold the cottage off. Three years later, the mill was put up for auction and several buyers were expected to make bids upward of £20,000. However, much to Miss Plesch's disappointment, Mr. Peter Benning succeeded with a bid of £5,500.[3]

Mr. Benning had the tower-mill renovated and with the aid of a grant from Brighton Corporation (part of which was later repaid) Vincent Pargeter, the present-day East Anglian millwright fitted new sweep frames in 1972. The tower was given several coats of white mastic paint, making a pleasing contrast to its appearance in 1971, when damaged sweeps and stocks were missing, as was the fan. The fanstage is now also in very good order, although not used for its original purpose.

Recently a large, well-planned modern extension was added to the tower to form an integrated dwelling, which, in early 1976, Mr. Benning sold for a sum in excess of £40,000 to Mr. and Mrs. Dowd.[3]

The best way to reach this site is to approach it through the estate of bungalows which flank the hill to the seaward. This will bring the motorist to a point within 50 yards of the now enclosed mill ground.

[TQ 292-086]

POLEGATE (Mockett's or Ovenden's Mill)

Restored Red Brick Tower-mill with White Cap and Fantail

We have travelled across mid-Sussex from behind one seaside County Borough to another, i.e., that of Eastbourne. Polegate village is about four miles inland and a major road and rail junction hereabouts. The mill, now restored by the Eastbourne and District Preservation Trust (Plate 82), has had a long and useful life compared to many in Sussex. Built in 1817 for Mr. Joseph Seymour in whose family it stayed for some 40 years. The next owner was Mathias Mockett who ran this windmill, Wannock water-mill, the *Red Lion* inn at Willingdon, and several other enterprises. Mr. Mockett gave his name to this mill, and to this day it is sometimes known as Mockett's mill. In 1918 Mr. Ephraim Ovenden came here after working at Cross in Hand and Dicker. One of a well-known Sussex milling family, he took on the business from George Thomas,[2] a descendant of Mr. Seymour. The mill was worked by Mr. Ovenden, and then by his son, Albert, up till 1965.[4]

When Hemming passed this way in the early 'thirties, the mill was established as Ovenden's (Plate 83). An appropriate name to be connected with country milling and baking. A contemporary photograph, of which there are many, for here was a popular subject, shows this mill, its granary and a malthouse standing at the field end of a lane which leads from the Eastbourne road. Today the site is completely surrounded by houses, making it perhaps a little out of place. Nevertheless, she stands proudly displaying her full compliment of partially-shuttered patent sweeps and a well-restored fanstage. Final weakening of this stage was the reason for the mill to stop grinding by wind during the last war. From then on, until Polegate mill finally ceased work eight years ago, it was powered by an electric motor within. This drove one pair of underdrift burrs, plus the cleaning and dressing machine. At the time of its retirement, the mill was generally rather dilapidated, the fan gone, its stage rotten, and the sweeps in a very weatherbeaten condition. The whole generally run down.

Fortunately for all lovers of such things, and of course future generations, Polegate was saved from possible demolition, by the Eastbourne and District Preservation Trust,[2] who were able to purchase the mill and its ground for £3,500 from Mr. Ovenden. The result of all these efforts, both volunteer and the work of E. Hole and Son, can now be viewed at leisure when the mill is open.

The fanstage is a facsimile of the original, and as well as carrying the ventilated five-bladed fan, is somewhat unusual in its two-tier construction, having two stage levels supported respectively by the main sheers and the fan posts. Tucked close to the petticoat of the cap, is the main cap drive wheel. This is worm-driven from the fan by a drive shaft with a crank in it for hand winding. This, we will remember, is a common feature as we have already seen at Icklesham and Cross in Hand. An unusual feature here is the weight wheel arrangement. Instead of being mounted vertically either to one side of or immediately behind the windshaft, the 'purchase shaft' is vertical and the wheel can be seen mounted horizontally above the cap rear. The weight chain has, of course, to pass over a pulley in order that it may hang to the gallery, which once surrounded the spout floor.

The shuttered patent sweeps are a reconstruction of the last set to work and still have some of the shutters in place with all the original striking gear (Plate 83). The sweeps are only allowed to turn when the wind is right, for the fan no longer winds the cap.

Inside the mill, nearly all the machinery survives. From the bin floor we can see the iron windshaft with its wooden brake wheel and brake, and at the tail end of the windshaft the rack and pinion for the weight wheel. Immediately above is the wallower and friction drive for the sack hoist. Below us is the stone floor which has three pairs of stones, one of peaks and two of burrs, the mill was originally fitted with two pairs of stones. One of the burrs is underdriven by auxiliary power (these were fitted by Medhurst in 1862 and were overdriven at that time), the other two being quant driven (overdrift) from the great spur wheel. On the spout floor, we find the tentering gear and the spouts plus a flour dresser and a 'Eureka' smutter by S. Howes of Silver Creek, New York.

For those who are further interested, there is available a very informative souvenir guide to this mill which can be obtained here.[5]

Here also is a very interesting mill museum with many exhibits and photographs of Sussex milling. Also a very fine working model of a Sussex post-mill, which was made by Mr. Frank Gregory. Mr. Gregory is a very active windmillian and has been very much involved in the work of restoration of many of the mills in this county. This, and the post-mill at Nutley in particular, owe a great deal to his dedicated volunteer work.[3]

The tower is of red brick, which was partially hung with tiles. Most of these have fallen off or been removed today. The spout floor doors open into space as the staging has gone, this having been supported at its outer edges with railway sleepers for many years, and has not as yet been replaced.

In July 1974 one of the sweeps fell without warning, breaking off at the canister, dragging its portion of the striking gear with it. The preservationists were quick to remove the complementary sweep in order to restore the balance. Needless to say, great disappointment was felt, as the stocks were only seven years old and should, all other things being equal, have hung for at least four times as long. The well-known Lincolnshire millwright, Thompson of Alford, has recently (May 1976) made and put up a new stock and two new sweeps at a cost of £2,000.

Take the A22 south from Polegate. The mill can be seen to the right about three-quarters of a mile on, up a short lane. [TQ 582-041]

SELSEY (Medmerry Mill)

Converted Red Brick Tower-mill with Fantail

This we could call a 'seaside' mill, for it stands about 200yds. from the shingle scarp which protects the flat hinterland of the Selsey peninsula from the encroachment of the sea. Coupled with its position, the mill is part of a large caravan site complex which includes all the facilities associated with a seaside holiday, and is itself, a gift shop and 'attraction', illuminated at night (Plate 87).

Various mills, both wind and tidal water, have worked here on Medmerry Marshes for over 250 years. The tower-mill that stands today, however, is somewhat younger. Various sources have put the date of erection at 1750, but a single word in a fire insurance schedule gives the lie to this:

> Mr William Reeves of Selsea in Sussex, Miller.
> On his windmill with going gears situated as aforesaid, timber: £200
> On his dwelling house only, distant, brick & tiled: £200.

'Timber' is the key word and obviously describes an earlier post- or smock-mill. The curious description of the house as 'distant' indicates a lessened risk of damage should the windmill catch fire. The date of the insurance was 3 December 1782.[1] During the early part of the 19th century, Medmerry mill was associated with a salt-panning business, an activity which was very profitable and much in evidence on the coastal plain of Sussex at that time. A notice of sale for 7 April 1827,[2] describes the business of John Reeves, various utensils used, and goes on to mention the mill as a 'substantial brick built freehold mill, carrying two pairs of stones with flour machines and joggling screens'.

We must conclude from this and the above that the tower-mill was built at sometime between 1783 and 1827. My feeling here is that she is a contemporary of East Wittering mill which was built *c.* 1805.

Work here, grinding for flour and feeds, continued until approximately the early 1890s. Millers[3] included H. R. Arnell, 1858-1878, Sampson Copstake, and F. W. Sharpe, 1882-1890. The mill then fell into disrepair and was taken on by Farne and Co. in 1905[3] or so. I believe that they attempted to run the mill for a time but finally decided to have her re-fitted. Holloways of Shoreham completely gutted the five-floor tower in 1907/8 and rebuilt her as a four-floor grist-mill.[4] She was equipped with a new iron windshaft and brake wheel which bears Holloway's mark (Plate 86) and a full set of broken applewood cogs today. Various cracking and kibbling machines plus a large oat-roller were fitted but no stones. A new cap and fanstage were built and the whole finished off with a full set of patent sweeps. Thus the mill worked until the early 1920s by Farnes[3] (Plate 86), when in spite of desperate attempts to keep the business going (such as the grinding of peppercorns, I believe) it was stopped. The evidence of the brake wheel indicates that she ran away in a storm at some time. A photograph, *c.* 1928, of Medmerry mill in *Sussex,* by S. P. B. Mais, shows each of the four patent sweeps to have lost the last five training edge bays, giving her a curious scarecrow look. The result no doubt of stormy winds which bend the backs of trees in this part of the county.

Latterly, the mill had a fine coppered beehive cap topped by a ball finial. The fantail and windshaft were serviced by an outside gallery, which ran round the cap's outside edge. A penthouse storm door opened out on to this at the fanstage. Today, however, the cap has been 'rationalised', and can no longer boast this feature which was probably unique in Sussex. The fanstage has also gone (removed in 1960) and large 'viewing' windows replace the storm door. Today the sweeps are dead and at this time the wind has left one of these damaged.

Like East Wittering, the tower is very well built and quite able to stand 170 years of so of Channel weather. As mentioned earlier here we are almost at the water's edge and the area is very flat. A little to the east is the village Selsey, famous for its lifeboat station and giving its name to this part of Sussex. We could almost be in another county, for the coastal plain is quite unlike its counterparts, the Weald high and low, and the eastern and western Sussex Downs. The land here is very productive, where a patchwork of market gardens and farms, some perhaps still growing the famous Chidham wheat, are the main source of livelihood.

From the high point of Halnaker Hill, we can view this peninsula to great advantage, and as with other mill hills, gain a better perspective of this county's beautiful variations, not always appreciated, even by those long lived here.

Selsey Bill has been eroded to some degree during the mill's life here. This is noticeable when comparing the first edition of the Ordnance Survey map with the present-day edition. We find, for instance, that the Medmerry Marshes around our present mill have been cut back, placing this mill about 200yds. nearer the tideline than when builders first laboured on the brick tower.

The area here has many historical connections. King Canute reigned from these parts and perhaps attempted his legendary turning of the tide here. Less widely known, it is thought that the Romans first landed in Britain in A.D. 43 along the western beaches of Selsey Isle, starting their long stay in Sussex.

We must take the A286 off the Chichester by-pass, then join the B2201 to Selsey, where we take a right turn to West Sands at which place Medmerry mill stands. **[SZ 844-934]**

STONE CROSS (Westham)

White Tower-mill with Fantail almost as it was when last worked

We come now to the final note on tower-mills, and indeed the last of all those Sussex mills standing in a 'proper' form, and I must admit my particular favourite of the tower-mills.

Stone Cross mill is the younger cousin of Cherry Clack smock-mill at Punnett's Town, 10 miles north of here. There are three points that encourage us to conclude this relationship. First, the mill was built for Mr. Samuel Dallaway, a cousin of the original owner at Punnett's Town. Second, the millwright was Stephen Neve of Warbleton, as he was for former. Third, we may regard the tower-mill as such kin to the smock-mill. Neves were the millwrights for most later smock-mills in East Sussex. The tower-mill here has a fanstage remarkably similar to the original stage of its cousin. The tower was contracted separately to Mr. Honeysett of Hurstmonceux in 1876[1] and on completion, was fitted out by Neves at a cost of £2,700.[1] I feel that there must have been a close working relationship between builder and millwright, for this mill is very well fitted out and its working proportions appear to be ideal. The mill here is one of the later tower-mills and its owner obviously wanted his mill to be as pleasing to look at as it was to work. As we can see, he must have been satisfied with the result.

Samuel Dallaway worked his mill until 1878, when his son, Frederick, took it on, working until 1895.[2] H. Dallaway, the grandson and the last of the family to work a windmill commercially, ran the business from then until 1937. It is interesting to note that his grinder apparently lived in at the mill house, for a set of back stairs was provided for use on windy nights (Plate 89).

The small brick tower, which houses five floors, is cement-faced and painted white. It is 16ft. 6ins. diameter inside the ground floor, the 2ft. 6ins. thick walls tapering upwards to 11ft. diameter at the curb. approximately 38ft.

above the ground. At each floor, a course of bricks is lost from the inside, thus forming a sill for the joists. An attractive feature, the windows, nine in all, are circular, though two of these are inside the roundel. The original apertures are filled round the metal window frames to complete the appearance (Plate 88).

A perfect beehive cap tops the mill tower this being covered with double-lapped ash boarding laid vertically and finished with a ball finial. Inside here, there are inscribed many names of the people who have visited the mill in the past. Most of these are of servicemen who were stationed here in the last war. The fanstage is served by a full-sized penthouse door opening inwards. There is also a 2ft.-square storm hatch in the front of the cap, which opens over the windshaft canister.

The windshaft is of iron and carries an 8ft. 7ins. diameter iron brakewheel with wooden morticed teeth. Cast in halves, the wheel was bolted together, thus clamping it to the square shoulder provided. Only two of the patent sweeps remain, complete with striking gear. At the back end of the windshaft, the striking rod passes through the tail thrust brass and is operated by a Kentish lever mechanism. This terminates below the fanstage with the weight lever and its chain. Originally, the striking weights were attached to the end of this.

The eight-blade fan winded the cap through bevel gears and a worm which engages with the iron rack. Mr. Hall, the present owner, has rebuilt the fanstage, renewing completely all the pitch-pine timbers. However, the eight fan blades have not been replaced as the cap cannot at present turn. The curb ring has been damaged and sunk a little. This is due to weight of cap, sweeps and windshaft totalling approximately nine tons, bearing down in the same position since 1937, when the mill stopped work. The cap is supported by rollers mounted in plummer blocks under the wooden cap ring, three under the windshaft bearing, and six others spaced around the curb.

Below is a gear floor proper. The iron wallower meshes with the brake wheel and an iron cone clutch for the sack hoist is neatly guarded with a rail.

The bins take up the next 6ft., these are totally enclosed and have a service door. As we pass down the steps, the great spur wheel appears, mounted above the stone floor. The stone nuts are disengaged now and the quant poles set aside at rest.

This floor is served by two of the pretty round windows and is light and airy. Three pairs of 4ft. ϕ overdrift stones, two of peaks and one of French burrs in their boxes, take up most of the floor space. The gearing for one of the peaks is the same as for the burrs which suggests that two pairs of burrs and one of peaks were originally fitted. Here also, is an Armfield dresser which is driven by an auxiliary nut from the great spur wheel. This floor is very well equipped and laid out, as indeed is the whole mill.

In the spout floor below are the remains of an oat-crusher which was belt-driven from the auxiliary shafting. A vent pipe in the tower wall here coped with the dusty efflux of a W. R. Dell smutter and cleaner, also fitted and driven similarly. The rest of the machinery of the spout floor is all in place. Three sets of tentering gear completely constructed of cast and wrought iron components await the motion of the stones above as the bins await the warm heavy meal. From the spout floor, we can either continue down into the noticeably dry ground floor, or step out on to the staging formed by the roundel roof. Here we can see close at hand the two remaining patent sweeps. The other pair were removed after storm damage in November 1928. We can also take in the view. Mr. Dallaway must have often stood here and gazed out to the sea two miles south, or towards the Downs behind Polegate, whose tower-mill can be seen not far west. These two tower-mills are now fixed to face each other across Willingdon and Mountney Levels. Perhaps Mr. Dallaway and his grinder timed their progress by the trains which take the rail spur here through Pevensey to Hastings. This spot is only just under 100ft. above the sea and yet commands the wind coming from any quarter, but especially the south-west, in which direction lies Eastbourne, whose wealthy conurbation is seen expanding rapidly across the Crumbles. Polegate and Stone Cross are the last of many mills to work in and around Eastbourne.

Mr. R. Hall was kind enough to allow me to spend many hours here. He hopes one day to have the mill opened for public viewing. At present he keeps the mill renovated amongst many other tasks.

The mill stands today as it was left in 1937, with two sweeps and no fan. Mr. Hall bought the whole property with the mill-house from the Dallaway family in 1966. The mill ground is a maze of outbuildings, amongst these is the miller's office still containing a lot of items dating from its use as such. The roundel houses many milling bygones such as sack scales, sack truck, steel yard, portions of sweeps, and a full set of shutters; also the fan blades. All these, it is hoped, will form a small museum.

Easily seen in Stone Cross village to the south of the A27 here.

[TQ 620-043]

CHAPTER SIX

NOTABLE REMAINS AND BYGONE MILLS (Fig. XI)

THIS CHAPTER is regrettably rather large. I say regrettably because it serves to emphasise the state of the county's windmills. Mentioned here, are 42 notable remains which, taken with the 28 structures covered in Chapter Five, gives a total of seventy. Any notable part of a windmill structure which remains above ground at the original site has been taken as worthy of mention here. Four very well-known bygone mills of which nothing remains, are also included, these being the famous six-sweep post-mill at Kingston, which fell in 1916, then Mr. Catt's elegant smock-mill at Hailsham, destroyed by fire in 1923, and, lastly, the well-known Navarino mills which stood in East Worthing until the early 1900s.

All the extant structures and notable remains are shown in Fig. XI, and the map references are given with each note. I hasten to add here that the reader will find a further list of 19 remains which are classed as obscure today on page 165. The grand total, including all of these, is eighty-nine. The list on page 182 shows these, and those shown on pages 170 and 177 will give the reader a further insight into the past.

I

POST-MILL REMAINS

ANGMERING (Luck's Mill)

Remains of Black Post-mill with Tail Pole and Talthur

The pretty village of Angmering has two main streets, that which runs east to west concerns us here and leads us, as it also will in the context of Highdown tower-mill, to the site of a rather illusive remain. Luck's post-mill, as the name suggests, was worked for many years by the Luck family (Plate 90) (1874 till 1930), first by Mr. Fredereick Luck, who went over to steam in 1899, and then with his son.[1] Here was a rather attractive black mill, enhanced by its ogee section roof. When working, prior to 1914, when she stopped for good, she carried patent sweeps which drove a pair of peaks and a pair of burrs. I believed that latterly a gas engine drove a steel disc mill in the roundhouse.[2] The whole was winded by tailpole and talthur.

By the time that Mr. Luck came here the mill had been working for at least 65 years, for it first appeared on the 1813 edition of the Ordnance Survey map. Previous millers included James Carter, 1839;[3] George Smith, 1845–58; and Peter West, 1858–74. The sweeps were removed in 1915/16 and by 1930 the remaining mill was fit only for demolition, and so in July of that year, the centre post was sawn through and the whole pulled down.[2]

Today, the top of the post can still be seen just showing through the roof of the roundhouse (Plate 91), which appears to have been a summerhouse or store. A large strengthening hoop is still holding the tarred brickwork together. This stands to the east of a pony field which is at the end of a short lane leading south off the village street mentioned earlier. **[TQ 072–041]**

APPLEDRAM (Dell Quay)

Converted Roundhouse of a Post-mill.

A miller insured his windmill here on the eastern banks of the Chichester Channel in 1790 for £200.[1] This is the earliest reference that we have of a mill in the parish. Two sale notices in the *Sussex Weekly Advertiser* during the years 1807[2] and 1810,[3] describe this as a 'large square built post mill, will grind 7 loads per week', also 'having 2 pairs of stones and a roundhouse, in the occupation of Mr. William Wild'. The first edition of the Ordnance Survey map 1813 does not, however, show this mill, which suggests that it may have gone by this date, and another erected by 1823/4, when Greenwood recorded it on his map of Sussex.

Dell Quay post-mill worked until sometime around 1870. The business was run by Farndell and Son of Dell Quay for at least 28 years[4] up till 1866,[5] when Mr. George Eastland, the last to work the mill, took over. There doesn't appear to be any further reference to a miller or mill in the parish after that time, and it is thought that the body was pulled down, leaving the roundhouse standing.[1] This was thatched and used as an outbuilding for the large house which stood a little to the south. The remain was integrated with 'Dell Quay House' (Plate 92) in 1948. The present owner, Mr. William Mason, was kind enough to show me the house and garden. Just to the north and seen in the illustration, is a swimming pool which marks the site of an old salt panning pond.

Dell Quay is reached by taking the No Through Road west from the A266 after passing southward through Appledram. This remain is easily seen from a catwalk that runs north from the quay itself. [SU 835-029]

ASHURST

Remains of a White Open Trestle Post-mill with a Tailpole and Talthur

Unlike many carved dates to be found within the windmills of this county, 1789 incised on the centre post here is proved beyond doubt by a sale notice in the *Sussex Weekly Advertiser* dated 13 July 1795:

> To be sold by auction by John Marwick on Friday the 24th day of July 1795, at the Red Lion Inn, in Ashurst Sussex. A new windmill in substantial repair, under a lease of 99 years, six years only expired, ground rent only five pounds a year. Now in full trade' enquire of John Younge of Ashurst.

Unfortunately all that remains 183 years on is the post, carrying the date, a peak runner stone, and a French burr bed stone. The burr is laid in the floor of a sun lounge on the southern side of 'Old Hill' house. High on the chimney work here is a small plaque which carries the inscription

BUILT BY
ML
1952

this commemorates the erection of the house and incorporation of bricks from the piers which supported the trestle of the post-mill.

Ashurst is a quiet low Weald village situated within four miles of the Sussex Downs and near the River Adur which has, as mentioned earlier, passed Shipley smock-mill four-and-a-half miles north-west. The village mill stood close to the inn on the slight rise in the meadowland. She was an attractive white open trestle post-mill winded by tailpole and talthur and standing on high brick piers. Latterly, her four patent sweeps drove a pair of peaks and a pair of burrs via an iron windshaft within (Plate 93).

F. Cooper, the millwright from nearby Henfield, was employed to fit a new clasp-arm iron brake wheel in 1864,[1] and is thought to have been responsible for making the change from spring to patent sweeps at that time.

By 1900, Ashurst post-mill was nearing the final decline, and whilst still grinding for flour and feeds at that time, closed down within a year or two. I believe that Mr. H. Barnes was working her then as a tenant of Mr. Etheridge. The Rapley family were early millers, first recorded here in 1845 and had the mill until 1858, when Daniel Chandler is shown as miller and grocer, followed by J. Duntries in 1866.[2]

After 1900 or so, decay was allowed to take its unhindered course, and by the 1920s (Plate 93) giving rise to concern amongst local people and various preservation authorities. Appeals were made to the owner to have her renovated, but these were of no avail.[3] By 1928 she stood minus a sweep, the other three badly battered, stripped of weatherboarding, her working gear exposed to the elements, and huge bramble bushes surrounding the substructure.[3] Then, in December of 1929, a fierce winter gale felled her, leaving Nutley, which had been shored up in 1928, as the only open trestle post-mill to survive in this county. Plate 94 shows the top of the centre post today, buried in the garden of 'Old Mill' house. As can be seen, this is crowned by a cast-iron samson head, the pintle appears to have been broken away during the fall. **[TQ 181-160]**

BEXHILL DOWN (Hoad's Mill)

Remains of White Post Mill with Tailpole and Talthur

Here are the remains of a fine hand-winded post-mill (Plate 68a), well described by Hemming. This, known as Hoad's mill, had two large pairs of overdrift stones (4ft. 6ins. Ø and 4ft. 3ins. Ø), and was last worked for provender in 1932.[1] The mill stood until 1965, when the whole body collapsed, leaving the trestle and roundhouse (Plate 95) as we see it today. The present Mr. C. Hoad (see page 144) still runs the family bakery business here and has a number of typical enamelled advertisement plates (associated with bygone milling) nailed to the side of one of the outbuildings. Also to be seen are the quants and iron windshaft (Plate 176). Nearby is the massive iron samson head bearing which supported the mill at the crown tree. This huge timber is elsewhere on the site, along with the remains of the side girts, stocks and sweeps. The samson head assembly may have been fitted later. Nevertheless, the mill was at times hard to bring into the wind, Mr. Hoad having to wedge his shoulders against the tailpole to heave the huge bulk round.

The post-mill was a very fine one built by a millwright called Plumley and documented 1784,[1] but Mr. Hoad believes it to have been older. Mr. Fuller, for whom the mill was built, was recorded here in 1820.[2] Later in 1839[3] Samuel Hayward, who worked the mill until Mr. Reeves took it on in 1856 or so, is shown for tithe apportionment. Mr. James Hoad worked as grinder for Mr. Reeves. Brothers James and Thomas Hoad then worked[1] for Mr. David Page from 1884, and took the mill on in 1887. Shortly after, Thomas went to work at East Hoathly on his own account as a baker, leaving James Hoad, who was the present Clifford Hoad's grandfather. The mill and business has been in the family for the 90 years since.[1]

Many of the improvements that came were included. Originally the mill had a wooden windshaft and spring sweeps. Later it was fitted with automatic or patent sweeps mounted on an iron windshaft, perhaps by Medhurst of Lewes, who had the licence for patent sweeps in Sussex and Kent. Upfields of Hastings looked after the well-being of the mill.[1]

Well worth a visit to anyone interested in the windmilling scene. To be found immediately above Bexhill Down north of the A259. [TQ 733-087]

CLAYTON (Duncton Mill)

Roundhouse of Post-mill with Tailpole and Talthur

Plate 71 shows the small roundhouse of Duncton mill close to the tower of Jack, the large tower-mill on Clayton Hill. The photograph shows the scene around 1905 when the miller, Mr. Guy,[1] was still working the mills.

Jack and Jill still stand today as described earlier and the roundhouse of Duncton mill is used as a study by the Longhurst family, who own the property.

Known variously as Duncton Gate, Dungate, and Duncton Mill, she was built in 1765 and leased for 99 years by Viscount Montague to Mr. Edward Oram of Clayton.[2] A series of sale notices in the *Sussex Weekly Advertiser*[3] show an early succession of millers: 1787, Mr. Horam; 1809, John Geere; 1810, Thomas Hicks; 1816, John Hamlin, who described the mill as a 'substantial built post mill carrying 2 pairs of stones'. She undoubtedly started life as a hand-winded mill with common sweeps. James Mitchell came some time later and was shown for tithes here in 1838.[4] *Kelly's Directories* show James Mitchell from 1845 onward. He, and his son (also James) worked the mill. His account book for 1849, quoted below, shows that business was good, and this no doubt prompted him to bring up a more modern post-mill from Brighton in 1851/2, later to be known as Jill. In 1866, two years after the lease had run out, it was decided to replace the older Duncton mill with a modern tower-mill, and so it was that Jack the tower-mill was built. The brakewheel from the old mill was used in the new, and was said to creak mournfully when under load in a stormy wind. Sadly, nothing remains of this today. A doorway between the ground floor to the tower-mill and the old roundhouse allowed the miller to use this as extra dry storage space. The later miller, Mr. C. E. Hammond, had an oat-rolling machine installed here.

To be found as for Jack and Jill. [TQ 304-134]

EXTRACTS FROM THE 1849 LEDGER KEPT BY JAMES MITCHELL OF DUNCTON MILL, CLAYTON. (Kindly lent by Mrs. Longhurst)

Charges for various products [these varied a little from one customer to another according to availability and means, I suspect]:

Wheat as:	per bag (3 bushels) £ s. d.	per bushel s. d.	½ stone measure s. d.
(firsts) Superfine flour	1 12 0	10 8	1 2
(seconds) Coarse flour	1 8 0	9 4	1 1
(thirds)	1 5 4		
sharps		2 6	
fine pollards		10 11	
coarse pollards		7½d.-9d.	
bran		7½d.	

Other products:

ground barley	3 9
ground peas	4 9

And for grinding customers' own produce:

wheat	8d.
oats	6d.
barley	4½d.

Extracts from actual accounts:

William Durtnall—

Date	Item	£	s.	d.
August 13th	2 bags of coarse flour	2	16	0
	8 bushels of coarse pollards @ 7½d.		5	0
	1 bushel of ground oats		3	0
	½ bushel of barley		1	10½
16th	1 bag of thirds flour	1	5	4
21st	2 half of ground oats			9
23rd	3 bags of coarse flour	4	4	0
	4 bushels of fine pollards		3	4

James Mitchell (his son)—

Date	Item	£	s.	d.
October 26th	4 half coarse flour		4	0
November 5th	4 half coarse flour		4	0
19th	Bill to William Durtnall	1	7	6
26th	1½ bushels malt and hops		12	8
December 1st	1 pig 13lb. 6ozs. @ 3/8	2	10	5
12th	4 half coarse flour		4	0
20th	4 half coarse flour		4	0
1850				
January 7th	4 half coarse flour		4	0
	1½ bushels malt and hops			
	J. Smiths bill for Beer and spirits	2	9	6
16th	Cash		17	5

This account was settled by credit as wages indicating that his son was working the mill with him.

The whole year brought Mr. Mitchell in over £2,500 which seems a good turnover for the period and no doubt encouraged him to improve output further by bringing up another post-mill.

CROSS IN HAND (Old Mill)

Roundhouse of a White Post-mill with a Tailpole and Talthur

The roundhouse that stands here today 100yds. south of the new mill, marks a traditional mill site. A deed in the possession of Mr. C. J. Newnham, the owner of Cross in Hand mills, gives us a 16th-century reference to a mill on Waldron Down, as it was known. The indentured parchment reads in clear Tudor handwriting:

> In the one and Fortieth year of Queen Elizabeth's reign: (1598) Let, all that land enclosed by a parcel of waste ground called Waldron Down—
> 6 pence per half hear to Mr. Putland for 999 years,—
> Piece parcel of land and windmill from Lord Pelham of Laughton Manor.

Later, a mill is shown on Budgen's map of Sussex 1724 and was worked by the present Mr. Newnham's great-great-grandfather, William, from 1777 until 1783 when he gave up to Mr. Thomas Saunders. Whilst he was here, the mill was badly damaged in 1790 when, struck by lightning, she suffered considerable shattering of the main timbers.[1] Subsequently, the mill was extensively repaired and offered to let in 1791 at three shillings per annum from Laughton Manor.[2]

The old mill, which was hand-winded and had two pairs of stones, worked right up till 1903 or so, latterly in company with the larger new mill which had arrived at Cross in Hand in 1855.[3] She had been taken on by William Kenward in 1854[3] and subsequently became part of the business of Newnham and Ashdown in 1888 after the Kenward family gave up.[1] Her usefulness ended when one of the stocks broke. Neves were called in for an estimate of the damage, and it was decided that she was generally too far gone to be profitably repaired.[4] The money was spent on building a power-mill, which was erected shortly after.

Some of the machinery was put to good use elsewhere when the mill was dismantled a few months after the accident. The two undamaged spring sweeps went to Dicker New mill and the tail stones to Blackboys post-mill.[4] Today, as may be seen in Plate 97, the original single-storey roundhouse is still standing just below the present mill and the power-mill building. The roof was made weathertight and the whole used as a store.

To be found as for Cross in Hand New mill. [TQ 558-217]

THE DICKER (Golden Cross) (Wicken's Mill)

Converted Roundhouse of fine White Post-mill with Sussex Fantackle

Locally known as Wicken's mill, after the family who ran the mill from 1858,[1] this is the remain of a white post-mill erected just prior to its appearance on Greenwood's map of 1823/4. The owner, Mr. Bourne, has a very attractive modern home and the roundhouse is blended well (Plate 98). A rather pleasant touch here is that the outer buttresses of the piers have been left tarred as they originally were. An old postcard shows the mill working around the turn of the century. It appears to have been a Medhurst mill with Sussex tailpole fan-tackle, a large body and patent sweeps, working two pairs of stones. In fact, Medhursts re-fitted this mill after damage in a great storm during 1867. Before this time, Golden Cross had a tailpole and talthur only.[2] The latter-day similarity between this and Cross in Hand new mill is striking (Plate 99).

The mill worked up till the outbreak of World War I, when the sweeps were taken down. The intention was to demolish completely at this time, but it, in fact, stood bereft of sweeps until 1919 when, with the assistance of Mr. Pearce, a local millwright and Mr. Thomas Ovenden of Lower Dicker mill, the body was finally pulled down.[3] The trestle is still intact inside the two-storey roundhouse, and the centre post now terminates just above the quarter bars. The roundhouse was used as a store for some years after the demise of the mill. Today it is used as a sitting-room with a bedroom above.

Golden Cross mill had an early association with a well-known Kentish milling family. For one John Russell is shown on a tithe map for 1839[4] as the occupant, Richard being the owner at that time. By 1845,[5] Richard (the father) was given as miller, and it is believed that John had gone to work at Cranbrook[6] with his brother George, who had in 1832 started the long association of this family with one of the finest smock-mills in England.

At Dicker, the surrounding area of the Low Weald is very flat and must have been good for wind, the mill making a prominent landmark as does the residence today.

To the south of *The Golden Cross* inn on the A22 Dicker to Lewes road.

[TQ 537-123]

THE DICKER (Lower Dicker New Mill)

Roundhouse remain of a large White Post-mill with a Tailpole and Talthur

Latterly known as Ovenden's New mill, this was originally called Dicker New mill to distinguish it from Willard's Old mill, which stood at Golden Cross, a mile away. First shown on a map dated 1813, New mill is thought to have been erected in the early 1800s and worked on the Low Weald here for 125 years or so.

New mill was being worked by a Mr. Elphick in 1815[1] and then passed to the Hide family who were recorded here from 1839[2] until 1907[3] (Plate 101), followed by George Dann. Then, in 1913,[3] Thomas and Ephraim Ovenden took over. The various owners ensured that the mill was well maintained, and she was metal-clad at some time, I suspect by Medhurst of Lewes. Golden Cross mill had the advantage of her with patent sweeps and Sussex tailpole fan-tackle, for New mill was hand-winded and worked two pairs of stones latterly driven by spring sweeps, two of these coming from Cross in Hand Old mill in 1904,[4] and two from Alfriston in 1908.[5] Thus she worked and was advertised in *The Miller* for 13 July 1908 as follows:

> For sale Dicker New Mill Chiddingly, a well fitted steam grist mill, 2 pairs of stones with or without 12hp undertype steam engine. A post windmill.

Mr. Benjamin Hide who worked the mill from 1882[3] till 1907 installed the auxiliary engine which was used to drive the stones and an oat grinder via a long belt. I understand that this was not entirely successful.

The Ovenden brothers ran the business for a number of years until, after Thomas's death, the mill was sold in 1924 to Mr. Harbert[5] who then worked the windmill and a power-mill nearby as 'Ovendens Dicker Mills'.[5] Ephraim had gone to Polegate mill a little earlier.

On 28 December 1929 the mill collapsed, perhaps weakened by the same gale that felled Ashurst mill three weeks or so earlier. The date 1825 was found on the samson head bracket after the fall which suggests that this may have been rebuilt or re-fitted at this time.

The windmill business and the power-mill were taken on by Stricklands of Hailsham.[1]

Today the granary building, which stood close to the windmill, forms part of a house. The remains of the roundhouse (Plate 103) sporting a shallow roof, have been utilised as a garage.

Situated at the end of a private road which runs southward off the A273 along the Dicker. [TQ 552-117]

HAREBEATING (Hailsham)

Remains of White Post-mill with Tailpole and Talthur

This originally stood three-quarters-of-a-mile away in Hailsham and was known as upper mill to distinguish it from Hamlin's smock-mill, which stood to the south-east of the town.

Incised in the post is the mark TF 1789, which leads us to believe that this timber was used in an earlier mill, as upper mill first appeared on a map dated 1823. Later, in December 1869, she was advertised for sale 'by private contact' 'to be moved', with two spring sweeps and two dead, two pairs of stones, plus smut and flour machines. Mr. George Weller took the mill and worked it at Harebeating after the removal. He was followed by Mr. Charles Hobden, 1876/1885; George Boniface, 1885/1894; and James Daws, in whose family the mill stayed until 1930.[2 and 3]

Here known as Harebeating post-mill (Plate 102), it stood for 64 years. The four spring sweeps with which she last worked were removed sometime during the First World War, and the whole became quite derelict. An auxiliary engine drove the mill via an external pulley for the latter part of its working life.

Inside the body, the stones were overdriven when working by wind. The iron brake wheel was fitted by Neves and carried their mark: STEPHEN NEVE 1876.[3] By 1930, the timber structure had become so exposed to the elements that collapse was inevitable, but it was not until the spring of 1934 that it finally came.[4] The stones in the breast fell through, followed the next day by the whole front of the mill, which could bear the weight of the windshaft no longer.

Today the two-storey brick roundhouse stands roofed with corrugated iron (Plate 100). The centre post with its samson head and part of the crown tree thrust aloft. Inside the roundhouse, which is used as a garden shed, the trestle remains intact on first-floor piers.

The present owner, Mr. Martyr, who lives in 'Old Mill House', was kind enough to show me the pick which was once used for stitching the peak stones. He is very fond of this remain and sad that the mill is no more.

To be found on the edge of Eastbourne Levels a quarter of a mile down Harebeating Lane, to the east of the A295 in Harebeating itself.

[TQ 596-105]

KINGSTON DOWN (Ashcombe Mill) (Lewes St. Annes)

Bygone White Six-sweep Post-mill with Tailpole and Talthur

This mill, alas! is no longer with us. There is, however, a trace of brick at its site at Ashcombe in the Parish of St. Anne's. This shows itself when the field has just been cultivated.

It is almost certain that Medhurst, the Lewes millwright was the builder of this mill, being very close at hand. Evidence suggests that Ashcombe mill was erected in 1832[1] and worked by Mr. John Weston until 1853,[2] and from then until 1888 by the Martin family.[3] I believe that the Martins were related to the miller who had Chailey South smock-mill at that time. The smock-mill there was at one time mistakenly thought to have had six sweeps.

The accompanying photograph here (Plate 104) is an extremely fine one and is taken from a glass plate which dates from the mill's heyday. We see that this was a six-sweep post-mill and indeed believed to be the only post-mill so equipped in England. At this time the mill is seen beautifully turned out. The patent sweep assembly was massive, mounted in a three-way iron canister and controlled by a three-legged spider iron. Each forked leg operated two sets of sail rods through respective rocking levers. The sweeps themselves were bolstered with clamp timbers in a similar fashion to the method used at Shipley smock-mill. This assembly must have been unusually heavy and with the brake wheel weighed about 11 tons. The neck bearing which supported the greater proportion of this load was at one time of glass, perhaps fitted in an attempt to keep the coefficient of friction as low as possible. Rex Wailes, in his book *The English Windmill,* tells us that this was later removed.

In spite of its appearance, this mill did not excel in any way. The great sweep assembly powered two pairs of underdrift stones in the breast,[4] I understand. These would have been spur-gear driven and the layout similar to that seen in Cross in Hand new mill today. It was noted for its apparent fast running, perhaps an illusion created after watching a conventional four-sweep mill at work. The mill was said (and it seems logical) to be able to work when others could not make use of the available wind. I imagine that this must have been potentially the most powerful mill built in Sussex, and impressive to watch when working (see page 31).

The site here is on open Sussex Downland and good for wind. Perhaps the miller shown here enjoyed his work in this mill, or was it troublesome? The heaving of the great tailpole round to bring the mill into the wind would have been no light task.

Mr. C. Saunders was the owner from 1888 or so until 1912 when the Sussex Archaeological Society took on this mill,[2] but sadly, Ashcombe mill, 'Old Six Sweeps', as it was fondly known, was blown down in a gale during March 1916. Several photographs taken in 1914 show her to have become very head sick and neglected. The windshaft with its three-way canister was removed from the wreckage and preserved by Mr. John Every of the Phoenix Iron Foundry at Lewes.[5] I believe that this was cut up for scrap shortly after Mr. Every's death in 1943.[6] According to the Gurney Wilson note books, the date 1832 was found on a main timber when the wreckage was cleared away during April 1916.

I am indebted to Mr. Edward Reeves of Lewes for his permission to reproduce the photograph here. This was taken by his grandfather, *c.* 1870.

The site is to the north of the Downland walk from Kingston Hollow to Lewes along Juggs Road. **[TQ 392-089]**

LEWES (Malling Mill)

Converted Roundhouse of a White Post-mill with a Tailpole and Talthur

The scene shown in Plate 105 is almost idyllic, and one which anyone who dabbles in windmill photography must admire. Mr. Frank Burtt of Ringmer, who photographed many mills around Brighton and Lewes during the early 1900s, took great care to ensure that the whole scene was recorded and we can see that his patience was rewarded here. Lewes and the Ouse Valley are shown cloaked in snow and overlooked by Malling post-mill sometime just prior to 1908. On 8 September of that year, a fire started within the mill,[1] and by the time that the fire brigade arrived, there was little that could be done. Malling mill stood on rising ground about 110ft. above the river, and the steam-driven fire pump could only be employed to raise water from a small reservoir which could not abate the flames. Plate 106 shows what remains of the roundhouse today. This is a residence comprising the circular brick and flint walls and an extension to the east. As can be seen, this has been roofed in and stands today in a small garden.

In its final form, Malling was a large white tailpole and talthur mill with large 1½-shuttered patent sweeps. The whole stood on a single storey of tarred brick and flint. The upper storey of the roundhouse which enclosed the trestle timbers, was walled with corrugated iron and painted white.

The site at Malling is a traditional one for windmills, and is recorded as such from 1625[2] onward. In appearance, the last to stand here was fairly modern, and it is obvious that this was rebuilt and re-fitted. The lower storey of the roundhouse is thought to be that of a predecessor, and that the whole mill was raised 6ft. or so above this at some time.

The building or re-fitting was almost certainly the work of Samuel Medhurst, and typical of his work at Golden Cross, Hurstmonceux, and Cross in Hand, carried out in the mid-19th century.

In 1810 Malling mill was worked by Mr. Henry Packham, who took on the business when his father became bankrupt.[3] She was worked by millers D. Morris, 1845; Richard Markwick, 1855–1878; and then by F. Stone, who employed a gas engine from 1900 or so. At that time, Mr. Castle Lever of Brighton was the owner and the business worked by Mr. Stone, a Lewes baker.

When writing of this mill,[3] the noted windmill historian, H. E. S. Simmons, pointed out the unfortunate record that Malling mill had for accidents. There are four relevant reports in the *Sussex Weekly Advertiser* early in the 19th century. The first, dated 5 September 1803, describes how the grinder here suffered a fractured skull when the hoisting tackle fell on him, and again in April 1804 when the mill came off worst, having 'Cast a swift on its meeting with a cow that was grazing near who suffered little or no hurt.' Unfortunately the victims in 1817 and 1821 came off rather badly. The last report for November 1821 tells us how the brother of the grinder, who had come over from Poynings to see him, became entangled with the machinery and had to have his hand amputated as a result. A lot of reports tell of such entanglements at various mills around the county, but the majority are, tragically, of a tenor illustrated by that of 3 January 1817:

> As a poor woman named Wood was returning home from the house of the miller of Malling, to shorten her passage, cut across an enclosed spot of ground on which the mill stands in a direct line towards the swifts, one of which struck her on the arm and shattered the bone to pieces and before she could get out of the way, the next met her and dreadfully fractured her collar bone and one of her thighs. She was conveyed home where she languished until Thursday and expired.

Perhaps the miller had his mill raised shortly after this time thus placing the sweeps out of harms way and better for the wind of course.

The converted roundhouse may be found today up a short turning on to Malling Hill, about 150yds. east of the A26 out of Lewes. [TQ 422-112]

NORTHIAM (High Park Mill)

Remains of a Large White Post-mill with Tailpost and Talthur

A windmill is first shown here, high above the Rother Levels, on Gream's map of Sussex dated 1795, but it is traditionally held that our present subject dates from 1800 or so when the earlier mill was burned down.[1]

This was a fine white post-mill which in her final form carried single-shuttered spring sweeps with a continuous windboard along their leading edges (Plate 108). Inside the three-floored body, two pairs of stones were driven by wooden head and tail wheels respectively. These had iron centres mounted on an iron windshaft. Today, the remains of two pairs of 4ft. peak stones are set into the ground between the brick piers of the roundhouse. A pair of 5ft. ϕ stones is said to have been used at one time. Perhaps these were French burrs, although nothing remains of them today. Plate 109 shows the piers now forming part of a pergola. The wall and upper floor were cleared several years ago. The mill itself was dismantled in 1949,[2] when the body became structurally insecure resulting from 30 years of idleness.

The first positive reference to a miller at High Park mill, for there were several in Northiam during the early part of the last century, is that for George Springett in 1840.[3] Spencer Ashenden then retained the business for 50 years or so, until 1899[4] when Mr. C. J. Banister of Millcorner mill, Northiam, took it on. During 1914, the mill stopped work and was not started again until 1918, when Mr. A. Goble,[5] who had previously worked for Mr. Banister at Staplecross smock-mill nearby, came to work here. The stocks and sweeps had become rotten during the years of idleness, and after a year of use one broke away, damaging the roof. Its opposite number was then taken down, and the other pair were removed shortly after. Mr. Banister was last recorded here in 1930.[4] The property sold in 1932.[1] The granary, which is known today as 'Black Barn', was converted into a house during the next year or so.

The mill-house and the other storage buildings were sold separately with the mill and today form part of the home of Mr. and Mrs. Bull.

[TQ 823-252]

RINGMER (Glyndebourne Mill)

Remains of White Post-mill with Tailpole and Talthur

This once pretty Downland mill collapsed with a resounding crash in 1925.[1] All that survives today is the trestle, though on closer examination, we find that only the post is original, being supported by 'reproduction' quarter bars. The cross trees have now gone and the assembly is set in concrete.

This little white post-mill was owned by the Glyndebourne Estate. The site is on the Downs just above the world-famous opera house built by the Christie family, and not far from Glynde itself, which is the true home of the equally famous Southdown sheep.

Accompanying this note, the two photographs, Plates 109 and 110, show first, the mill when working as a tailpole and talthur mill with spring sweeps, *c.* 1910, second, the trestle as it stands today, looking towards Ashdown Forest. The top of the original post carries a modern plate, and above the lower iron hoop has the clear carved inscription LH 1706.

Glyndebourne mill is believed to be the one shown on 'Budgens' map of Sussex, 1724. The carving on the post may be that of the millwright. It is possible, however, as with many others, that this was rebuilt at some time, using some of the original timbers. She would no doubt have started life with commons, but ended up carrying one-and-a-half shuttered spring sweeps. These drove a pair of peaks and pairs of burrs. Thus the mill was worked until 1921.[2] Two years later, one of the side girts failed, causing her to take on a dangerous set, and final collapse at 4 p.m. on 6 June, 1925. The noise of this on a still summer afternoon, as can be imagined, caused great alarm amongst the local inhabitants who likened it to a crack of thunder. The mill was last worked by Mr. A. J. Holder,[2] who had been here since 1912, and followed T. Hollingdale. An earlier record of a tenant here shows that the fair sex was equally capable of running a windmill, for a lady miller, Jane Smith, was liable for tithes in 1843.[3] She was obviously successful, and worked here for 12 years, when Mr. John Smith took on the business.

Alfred Shaw, the millwright who continued the work of Medhursts when they finished in 1878, left his mark on the brake wheel.[1] For many years, the original trestle timbers stood, but eventually, like the mill, they fell. Several of these still lie in a small depression nearby. Fortunately, the spot is marked today as a tribute to this local landmark. The post was re-erected in 1968 by the Glyndebourne Estate Trust.

Easily found at this attractive spot by taking the turning to Glyndebourne off the A265 Ringmer to Lewes Road. **[TQ 450-112]**

RISEDEN (Wadhurst)

Remains of a White Post-mill with Tailpole and Talthur

In the garden of the mill-house at Riseden, a High Weald village in the north-east of the county, we find this remain. Here, until 1910, stood a white post-mill with a tailpole and talthur. This was latterly equipped with two pairs of stones driven by four patent sweeps.

First shown on Greenwood's map of Sussex dated 1823/4, this is traditionally thought to have been brought here from Bestbeech Hill where it appeared on Gardener's map dated 1795. But, although a mill at 'Bestbridge' Hill is mentioned in the 1801 Defence Schedule,[1] no mill is shown at either Bestbeech or Riseden on the 1813 Ordnance Survey map. The first reference to a windmill at the present site appears in the *Sussex Weekly Advertiser* dated 12 March 1821, and reads: 'To be sold by private contract, a substantial windmill with machinery complete, situated at Riseden near Best Beech Hill'. Millers included N. and J. Packham,[2] 1839; T. Martin, 1851–1874;[3] and then George White until the mill ceased working for a time around 1890. During Mr. White's tenancy, the mill was converted from two spring and two common sweeps to all patents.[4] The house and mill were taken on by the local vicar, the Reverend Roberts, who hired the mill to a Mr. Harry Wallis for a time.[4] By 1910, Riseden mill (Plate 111) had become idle and weakened to the point where it collapsed without warning on a still day.[5] The trestle stood, still supported by the roundhouse for some 60 years, until this also fell. A cross tree apparently broke away at a pier and brought the post down across the roundhouse wall, which crumbled beneath the weight. Today, the top half of the post, which was subsequently sawn off, stands in Mr. Overton's colourful garden and still exhibits the thrust-ring which once bore the weight of this windmill. Plate 112 shows the remaining roundhouse walls. [TQ 621-303]

SULLINGTON (Warren Mill) (Crowhurst's Mill)

Windshaft of Open Trestle Post-mill with a Tailpole and Talthur

Sullington Warren is an attractive upthrust of greensand half-a-mile east of Storrington. From the gorse-covered vantage point here, the view to be had of the South Downs and surrounding countryside justifies the patience involved in tracking this mill remain down. On gaining the south-east brow of the warren, part of which is National Trust, the walker will find a large cast-iron windshaft which to the windmillian is unique in Sussex today. As mentioned in the note concerning Climping smock-mill, this is one of two iron windshafts to be found which have two sets of cast canisters, one for the stocks and the other for the spokes of a compass-arm brake wheel (Plate 113). I suspect that the reason for using this unusual mounting arrangement was that this little post-mill worked both her pairs of stones via a spur-gear drive taken from the brake wheel.

Like the post-mills of Ashurst and Nutley, this represented the basic form of the windmill. She, in fact, worked throughout her life as an open trestle-mill with common sweeps and winded by tailpole and talthur. Shown on the first edition of the Ordnance Survey map, the mill is thought to have been built in 1800,[1] and took its name from the Crowhurst family who were millers here, and at a nearby water-mill, for many years. Mr. Robert Crowhurst had the business from 1851 until his death in 1870.[1] His widow continued until 1907[2] when the mill closed down. The stones and most of the machinery were removed by a local millwright and transferred to the water-mill.[1]

It was found in 1908 that the windmill was in danger of falling. She was treated to emergency repairs, including a pair of shoring poles under the breast.[3] By 1910 or so (Plate 114) the body and sweeps had become rather tattered, and the mill appeared to be awaiting the end. This came in August 1911, when the warren caught fire.[3] Strenuous efforts were made to deflect the blaze and thus protect the wooden structure, but it was a hopeless task. The iron windshaft survived to mark the site of this attractive little white mill to this day. This corner of the Warren is presently owned by the Horsham district Council. [TQ 097-143]

WINCHELSEA (St. Leonard's Mill)

Externally Renovated Black Post-mill Remain having had Top Fan and Trolley Steps

Here, over in the east of the county once more, we find this mill remain overlooking the site of ancient Iham, on the outskirts of Winchelsea. A walk across the cricket field on the west of the town, then along an attractive footpath brings us to St. Leonard's mill. This local name is traditional,[1] perhaps given by the miller who worked a mill nearby from 1703.[2] I think it is more likely, however, that this derives from the fact that beneath the roundhouse are the remains of the old church of St. Leonard's. This was, perhaps, the one-time ecclesiastic centre of the small parish of Iham which is shown within the larger parish of Winchelsea itself on the tithe map of 1842.[3] Approaching along the path mentioned, we seem to follow a ditch which was part of the 14th-century fortification of the town. This mill stands at the end of an earthbank which terminates in the church mount, part of the same earthworks. Cattle and sheep roam freely in respective areas of this beautiful setting (Plate 115). The mill is bereft of sweeps and windshaft but still retains the wooden brake wheel and remains of the tail wheel and stones; an iron girder serves as support for the two wheels today. This was very attractive when at work around 78 years ago (Plate 116).

Like Hog Hill mill, Icklesham, which can be seen from here, St. Leonard's mill had a top fan, and the drive to the step wheels may still be seen (Plate 115). Two spring and two common sweeps drove three pairs of stones in an unusual manner within. A pair of burrs and a pair of peaks were driven conventionally in the head and tail of the stone floor. The third pair of 3ft. peaks (still within) were underdrift in back of the bin floor, driven by a belt from an upright shaft which in turn meshed with the tail wheel.

The mill that we see today was either built here on the church mound or moved to this spot from the traditional mill site a little to the north, between the years 1813, when it was still shown at the old site, and 1823. It is thought that she may, in fact, date back to 1760[2] if the latter be true.[4] She was worked continuously by the Sharps family for over 60 years, not, however, without mishap, as noted in a contemporary news item for 27 August 1861.[5] This describes the inquest on the death of Benjamin King—millwright. He was, with the assistance of Mr. J. Sharps (the owner) and Mr. J. Whitman (loader and grinder), hauling up a new stock when the accident occurred. Mr. Whitman told the court: 'We were drawing up a large timber called a middling by rope when Mr. King came to assist and it broke, the timber striking him on the head'. The next statement seems tragically obvious to us now: 'It knocked him down and he did not speak afterwards'. The stock went on to upset a trestle that Mr. Sharps was standing on and he suffered a broken thigh as a result. The mill was subsequently repaired and worked until just before the turn of the century,[7] when it became idle and by the 1920s, quite derelict. During 1935, the whole was renovated at the expense of G. M. Freeman, K.C. In the meantime, however, the wooden windshaft had been removed and a small set of dummy sweep frames put up. These did not survive. The body and roundhouse, which had become badly weather-breached, were completely re-boarded and made good in 1955 by Dibleys, a local firm experienced in this type of work.[2] The mill has remained pretty well in this condition ever since, with periodic attention, including some shoring-up in 1961.[6]

In February 1978, sadly, gale force winds blew the back out completely from the steps to the eaves. She is set over a little and leaning back, and may be seen, a beam dislodged and one side bellied out somewhat. I understand that the National Trust, who have been custodians for the last three years, are considering some sort of restoration programme.

Is is worth noting here that St. Leonard's mill retained an early windshaft with a wooden poll end. This, a continuation of the wooden shaft, carried the stocks in mortices cut through the large square head. She was, I understand, the last windmill in Sussex to boast this feature.

If we approach this site from Icklesham mill, one-and-a-half miles away, we pass through the land gate mentioned in that context and find ourselves joining the A250. We must turn off this to gain a parallel street on the west side of the village. From here, the short walk described above can be taken.

[TQ 902-176]

II

SMOCK-MILL REMAINS

BALDSLOW (The Harrow Mill)

Converted White Smock-mill with a Fantail

Although this is now only the brick base plus two floors of the wooden smock-tower, it forms a pleasantly attractive part of a bungalow, which at one time was a granary and power-mill, today adorned with wisteria and surrounded by a very colourful garden (Plate 118).

Still recognisable as part of a windmill, it stands high above the village of Baldslow, three miles to the north of Hastings. Mr. and Mrs. Greenall were kind enough to show me the inside of the base which still has the huge crossing joists in its ceiling supporting the spout floor, now a bedroom, above. This octagonal ground floor is some 20ft. across, and whilst the furnishing must have been very difficult, the result is a pleasure to behold.

Harrow mill was erected here, on the site of Hayward's post-mill, by Upfields of Catesfield.[3] The old mill was pulled down in 1855[2] and the new mill put up shortly afterwards. She was a fine example of a five-floored white Kentish smock-mill with patent sweeps, hooded white cap and fanstage. This must have been a striking landmark for miles around (Plate 117).

The Hayward family owned and worked her for many years until the whole was sold to Colonel Hanky in 1889.[3] I believe that the Haywards continued as tenants for some time.

A late March gale in 1916 tore the cap and sweeps off, completely wrecking them, and causing considerable damage to a nearby store.[4] I understand that she had been allowed to run away after becoming tail-winded. The smock was patched up and the whole run as a power-mill until the early 1920s, when Mr. F. Richmond purchased the mill and granary buildings. He then made the conversion to a bungalow dwelling, incorporating the roofed-in lower three floors of the smock.

East of the A21 and A2100 junction. [TQ 801-132]

The above is one of the last remnants of many windmills around Hastings. The saddest loss in the more recent memory of Sussex milling was that of Silver mill or Draper's mill, which Hemming recorded at length. There, on the west of the town, stood this very fine smock-mill of the Kentish type, also built by Upfield in 1868. After strenuous efforts to save her, she was demolished in 1966. Today, a small industrial estate marks the site.

BILLINGSHURST (Hammond's Mill)

Derelict Stone Base of a Smock-mill, with Hand Winding Wheel

As is seen in Plate 119, this remain is almost completely engulfed by natural vegetation and is easily mistaken for a large bush. Here, in a field just east of the village, stands this two-storey base roughly built of local sandstone. The floor level is still discernable through the doorway, and we can see that a tree is firmly established now, growing through its middle. In R. Thurston Hopkin's book, *Old English Mills and Inns, 1927*, we find this remain sketched for us. At that time, the base was in much better shape and still boasted an inset tablet of stone which bore the clear inscription $\frac{\text{J.S.}}{1825}$.

Hammond's mill (named after the farm nearby) was built for a Mr. Richard Chennell by J. Streeter, and subsequently leased from July 1825 until September 1832.[1] Later in 1839,[2] a Mr. William Sprinks is recorded here; and again in 1841,[3] when this is referred to as 'East Mill'. Mr. Sprinks, who also worked another mill in the village, I believe, ran the mill until 1874,[4] when Mr. H. Isted continued, to be followed by Mr. W. F. Weller, 1878. In 1895 the sweeps were badly damaged whilst in the charge of an apprentice to a Mr. Underwood.[5] Apparently the hand-winded hooded cap remained until 1906, when this was blown off, leaving the mill capless and derelict for some years. By 1920 the smock had gone and the base demoted to a crude store. An old resident of Billinghurst still recalls playing in the base as a boy,[6] and it is said that the remaining smock was finally burned down when it became dangerously unsafe.

Clearly seen to the north of the A272 on entering Billingshurst from Coolham. [TQ 092-261]

CHAILEY SOUTH

Derelict Base of a White Smock Mill with a Fantail

Inside the derelict shell of this single-storey brick base is carved a datestone: I WICKER 1813, which first suggested to me that the mill was built here in that year. A mill was shown at Yokehurst, half a mile south-east of here on the first Ordnance Survey map and the present one first appeared on Greenwood's map of 1823 (Plate 120).

On contacting Mrs. Mutter of Mill House nearby, I learned that the mill was said to have been moved to the present site from Yokehurst, this from an old inhabitant who had been told of it by a predecessor.

The *Sussex Weekly Advertiser* bears out local belief and shows how misleading some datemarks can in fact be:

> September 18th 1808 'To be sold on 8th October next A large good timbered smock windmill adjoining Chailey South Common—to be removed—before next May.

This obviously referred to Yokehurst mill which had stood for some time as shown by an earlier notice dated 30 August 1790:

> Wanted a miller at Michaelmas next a sober steady man as grinder to a windmill at Chailey South Common—none need apply unless he is of good character and thorough master of his business. NB It is a new erected mill.

and so it would appear that the smock-mill was built at Yokehurst in 1790 or a little earlier and removed to the present site in 1808/9, too late to be recorded on the first Ordnance Survey map at the new position. Mr. Nicholas Olive was the owner in 1790 and the datestone may well be the mark of a miller in 1813 or commemorated raising the mill. The Hamshar family came soon after and were here until 1862, when Henry Hamshar died.

This white smock-mill with a fantail started life with commons. Later converted to two springs and two commons, and finished its working life with two patent sweeps driving a pair of burrs and a pair of peak stones. This was around 1887. The mill had been damaged in 1884 by a gale resulting in the removal of two of the sweeps.[2] Most of the machinery was taken out by 1890 and the shell then became derelict. A local lady[6] can remember well the mill standing here in 1906 beside Martin's bakery. Mr. J. R. Martin came here in 1866[2] and the family continued until 1900 when Mr. F. B. Avery took on the business, keeping Martin's name.

It has often been suggested that this, like Kingston Down post-mill, had six sweeps. This is thought to have originated from a notice of auction in 1862 when the mill was advertised with '2 new spring sweeps and 1 pair of dead',[3] mistakenly taken as two pair spring and one pair dead. In any case, a very poor postcard shows this derelict smock-mill *c.* 1900[4] still carrying a four-way iron windshaft. We can see in the illustration here just how stout the brickwork is, fully three courses thick with eight tapering buttresses three courses thick at the base. For a number of years, part of the wooden smock tower remained, and the whole was used as a store.[5]

Just off the Lewes to Chailey road A275 in the village of South Common.

[TQ 388-176]

CLIMPING

Converted Tower of once Black Smock-mill with Fantail

Climping or Clymping mill, as I think it should traditionally be spelt, was originally part of the Ballifscourt Estate and should not be confused with the tower-mill which at one time stood three-quarters-of-a-mile away on the east quay at Littlehampton.

This smock-mill is believed to have been built to replace an earlier mill in 1799,[1] which was to be found at one time carved on the beam inside, and stopped work around 1900.[2] It had a bin, stone, and spout floors mounted on a single-storey base built of knapped flints, brick and stone (Plate 121). A black beehive cap was winded by a fantail and carried spring sweeps, the whole working two pairs of stones. Tenant millers included the Barnard family from the early 1800s[2] till 1878, followed by Robert Coote until 1890, and Reuben Smart in 1895.[3]

After the mill stopped work she was allowed to become weatherworn and by the time Sir Richard Garton took on the property,[2] was rather derelict. Once again the remains of the smock and base are part of a house (Plate 122). The living accommodation was added in the 1920s when the conversion was initially carried out. The smock-tower with its cap and a full set of rather short 'mock' sweeps, stood up until 1963.[4] The whole of the weather-beaten upper structure of the mill had been found in 1962[5] to be infested with death-watch beetle. The only reasonable solution in the circumstances was to dismantle the top of the mill. An interesting point arose when the windshaft with its wooden brake wheel was removed.[6] This turned out to be one of two Sussex iron windshafts to have cast box mortices through its shank for a compass-arm brake wheel (see page 22). Older wooden shafts were often morticed in this way. This is now at the Science Museum at South Kensington in London.

The easiest way, and the most pleasant approach to the mill, I found, is to travel down the 'rope walk' on the west side of Littlehampton Harbour. Here is the mouth of one of the five main Sussex rivers, the Arun, which is tidal and flows out at great speed. We must travel to the seashore and walk three-quarters-of-a-mile towards Climping. This is Littlehampton West Beach. Here, incidentally, is what was the best piece of sandy shoreline in Sussex, when not too crowded. The house, which is today used by the Post Office Pension Fund Trustees, as a holiday home, stands inside a copse, about 200yds. from the sea, no doubt planted in latter years to protect the area from south-west gales.

Very much hidden in its surrounding copse, the house and mill remain stand almost on the shoreline here behind the dunes west of the golf course. [TQ 015-012]

COWBEECH (Hurstmonceux) (Erry's mill)

Converted Two-storey Base of a Black Smock-mill with a Fantail

This is one of these remains that can be rather difficult to find. Whilst it is actually in the parish of Hurstmonceux, Cowbeech is, in fact, two miles north of that village. The converted two-storey base has been given a tiled conical roof, thoroughly re-rendered with cement, and fitted with metal-framed windows. No doubt this is as comfortable a home as it appears to be from outside (Plate 124).

Here was a large six-floored mill, whose smock was covered in sheet iron and tarred black. Above this, the white-hooded cap carried a Kentish fanstage and four one-and-a-half-shuttered patent sweeps (Plate 123). The smock was painted white before being covered in iron, I believe. Greenwood's map of Sussex is the first to show Cowbeech mill in 1823, and it is believed that she was erected in 1820 for Mr. Brett.[2] Mr. Brett was registered here until 1839,[2] then Samuel Dallaway until 1878.[3] During his period of working the business he went over to steam when a Cornish boiler was acquired in 1860.[1] I understand that Mr. David Catt worked here in the 1870s before going on to Ocklynge mill at Eastbourne.[4] Mr. Catt subsequently went to work for Mr. Mercer at Hamlin's smock-mill at Hailsham (see page 116). From 1880 or so, the Erry family were millers at Cowbeech,[3] and in 1900 *The Miller*[5] carried the following notice:

> To be sold at auction at the Crown Inn Hailsham on June 27th. A well placed smock corn mill known as 'Cowbeech Mill' built of brick and timber (cased with sheet iron) and six floors with engine house, boiler, store room, van shed etc. The whole in the occupation of Mr. Joseph Erry whose tenancy expires at Michaelmas next.

At that time, the mill was equipped with two pairs of peaks and two of burrs. Two pairs were steam-driven, and two by wind. Another member of the family, Mr. Alfred Erry, took on the mill and is registered from 1903 until 1930,[3] a little later Sticklands of Hailsham bought the business. By this time, unfortunately, the windmill had been damaged by fire after being struck by lightning in 1911. The remains of the smock was removed just after the First World War,[1] and the base used as a store for many years, being converted into living accommodation during 1967.

A short way along the Bodle Street lane out of Cowbeech village.

[TQ 621-146]

CUCKFIELD (Beech Farm Mill)

Brick Base of a Slender White Smock-mill with Fantail

At Beech Farm a quarter-of-a-mile south of Whiteman's Green, stands this rather obscure remain. The buildings here are thought to have been erected in 1873[1] or so for the Sergison Estates, and it is almost certain that this mill was built at the same time to replace an earlier one which stood a little to the north of this site.

The photograph here (Plate 125), which is from the Gurney Wilson Collection, shows the mill *c.* 1910 as a little white lady with four short single-shuttered sweeps (one of which is apparently missing) and a hooded cap which was winded by a five-blade fan. The narrow octagonal smock was held to the base by stout tie-rods which in turn were anchored in large lug-irons on each side of the door.

The two-storey base survives today as an integral part of the adjacent barn. As may be seen (Plate 124), the walls are slightly battered and still bear the marks left by the tie-bars. Inside the first floor there is still evidence of the machinery driven here. Line shafting that once passed into the barn, remains. A winnowing cupboard also survives and gaps in the whitewashing of the wall space indicate various chutes. Like West Blatchington, this appears to have been a utility mill, perhaps used for driving farm machinery as well as stones, although there is no positive evidence of this today. There is, however, a modern oat-crusher powered by electricity in the ground floor. During the First World War, the mill became idle and by 1918 had sustained considerable damage to her sweeps.[1] I believe that the smock was dismantled in 1922 by a Mr. Morley, whose grandfather had built her,[1] and the base roofed in rather neatly. The whole property was bought at auction four years ago by the Mid-Sussex District Council, and I understand that plans are afoot to re-develop the land for recreational purposes. At present the farm is used for calf rearing.

To be found on rising ground just to the west of a footpath through to Cuckfield village. [TQ 297-253]

DALLINGTON

Remains of a White Smock-mill with a Fantail

Very little is left to mark the site of the white fantail smock-mill (Plate 128) that stood above Dallington. In a field just 200 yards north-east of the village church, stands one side of the octagonal two-storey brick base (Plate 127). The photograph also shows Mr. Dallaway's mill at Punnett's Town, two-and-a-quarter miles to the north-west. The surrounding area in this part of East Sussex is a panorama of three-lined ridges, which once boasted a number of smock mills. Within six miles or so of the Dallington site stood those at Cowbeech, Chapman's Town, and Lunsford's Cross. Two more, those at Battle and Punnett's Town, still survive today in recognisable form. Dallington smock-mill was built in 1851/2[1] for Mr. Albert Geering. The land was leased from the Ashburnham Estate, and Upfields of Warbleton[2] were employed to erect the white wooden smock on a tarred two-storey brick base.

A sale notice in the *Sussex Weekly Advertiser* for 2 August 1881 gives this as 'An exceedingly well built smock mill, situated near Dallington Street'. 'It is in good order and condition, built of the best materials, driving 2 pairs of stones, fitted with a grain cracking mill, the usual machinery and has patent sweeps plus brick and tile stables for thee horses and a van lodge'.

The foundations of the stables, etc., are still to be found close to the remaining one-eighth part of the base which formed an end wall.

Mr. A. Geering was registered here from 1851 till 1878[3] when George Geering took it on until 1895.[3] The lease was put up for sale in 1902,[4] as 'Dallington Wind & Steam Mill with stables and store on a 99 year lease from 29/9/1852'. The tenant at that time was Mr. Henry Wood. From 1903[3] Mr. William Richardson is registered, his grinder being a Mr. Traeton Booth. No miller was shown after 1905. The mill was working three pairs of stones, 4ft., 4ft. 4ins., and 4ft. 6ins. ϕ respectively, and an oat-crusher. I believe that one pair of stones and the crusher were auxiliary power driven.[4] She obviously deteriorated rather rapidly during the latter years and had the reputation of not winding well. Sometimes during late 1913, the smock was pulled down, leaving the base and buildings standing. Mr. Archie Dallaway remembers the brickwork of the base being mutilated and taken away for use elsewhere in 1924.

The present Ordnance Survey map shows this rather stark sentinel as a minute black rectangle and it is easily found in the centre of a field close to Dallington church to the south of the B2096. [TQ 660-193]

HAILSHAM (Hamlin's Smock Mill)

Bygone White Smock-mill with a Fantail

This is included as a notable bygone for there is nothing left of this fine white smock-mill today. Hubbard of Lewes erected her in 1834[1] and it was the only certain evidence of this millwright's work standing during the last 60 years (Plate 129). This was originally called Lower mill to distinguish it from Hailsham Upper mill (see page 100).

As can be seen in the photograph taken some years before disaster overcame her, she was equipped with double-shuttered patent sweeps operated by a weight wheel and chain. An attractively-sloped fanstage placed the large eight-blade fan well for wind. This was carried by a beehive cap, which must surely have inspired Neves when they built Stone Cross tower-mill on the next ridge south in 1876, the similarity of the cap design is striking.

Mr. Robert Kenward was the first to work 'Lower Mill' in 1834. A number of millers followed, and in 1859 Mr. John Mercer took it on. Messrs. Dann, Catt and Boniface were his manager, grinder and loader respectively, and when he retired in 1879, took on the mill as partners. During the next seven years, Mr. Dann went into a bakery business on his own account at Horshoe Green, Edenbridge, and Mr. Boniface went to Harebeating as miller, leaving Mr. David Catt, who set up as David Catt, miller and baker, in 1886.

The mill was re-named Hamlin's smock-mill after the farm nearby and latterly worked for provender only. She was equipped with three pairs of wind-driven stones, two pairs of burrs, and one of peaks. A fourth pair of composition stones were fitted in the base and driven by a 28/32 h.p. suction gas engine. Mr. G. A. Catt (see page 140), who is now 73 years old, worked in the windmill as a boy. His father, Ebenezer Catt, was by this time running the business as D. Catt and Son. The miller, Mr. Tingley, worked the power-drive stones and 'employed' Mr. G. A. Catt when the wind was good to grind farmers' corn. This was a general term for oats, barley, beans, and maize for provender.

Unfortunately, a disastrous fire destroyed Hamlin's mill in November 1923, and Neves were employed to fit out a new power-mill for Ebenezer Catt during the following year at a cost of £600. This worked two pairs of 4ft.ϕ composition stones and allied equipment driven originally by a 55 h.p. Tangye horizontal oil engine. Mr. Gilbert Catt finally closed down in 1967 and the power-mill was let out to a sack repair firm. Shortly after this another fire reduced the building to a mass of rubble and twisted metal. A sad end.

[TQ 598–089]

HUNSTON (Kipson Bank Mill)

Base of a Ten-sided Black Smock-mill with a Fantail

This mill formed part of the business of the Hodson family from 1865 until 1915.[1] The Hodsons were well known as millers during the latter half of the 19th century. A large 12-sided smock-mill that stood above old Brighton on West Hill was worked by them before being demolished in 1866. Kipson's Bank smock-mill was built in 1801[2] and first shown on the Ordnance Survey map of 1813. This was a large 10-sided smock-mill with a beehive cap of very similar design to that of Somerley smock-mill, three miles south-west of here. From this likeness and the closeness of dates of building, we may conclude that these two were the work of the same millwright. Two pairs of French burrs were powered by shuttered sweeps. The *Sussex Advertiser* for 12 March 1832 carried the following sale notice:

> Kipstons Bank Windmill, Hunston near Chichester, particulars of an excellent freehold smock windmill in full trade, to be sold by Henry Salter at the Anchor Inn Chichester, Tuesday 20th March 1932. The mill is worked by two pairs of French stones and all tackle and fabric in good repair. The exterior of which has been newly canvassed and pitched. Further particulars may be known of Mr Edmund Dash, Slindon.

By 1845[3] Mr. John Paice was the registered miller, followed in 1847 by one Alfred Millyard and William Dendy,[4] who worked the mill until the Hodsons bought the mill in 1856.[1] By that time the mill had been refitted to work two pairs of peak stones and one of French burrs.[5] A power-mill was, latterly used here and the buildings were associated with this still stand today close to the remaining windmill base. Mr. Hodson closed down in 1915[1] and the mill which had become derelict through many years of idleness (Plate 130) was pulled down in 1919. The business was sold to Bartholomews of Chichester.

The unique 10-sided base is 23ft. across corners inside, built of brick, cement faced and tarred on the outside.This was roofed in and used as a summer house. Today it is open to the sky and completely overshadowed by a large cedar tree (Plate 131), three of the stage supports remain, consisting of iron poles set in stone blocks at ground level, each supporting one of the 10 main spars. Elsewhere are five sections of the wooden curb and three portions of the wooden brake. Close to this are the other buildings including the derelict power mill, all of which are on private ground.

A large mill-house on the eastern side of the Hunston to Sidlesham road, B2145, pinpoints the site a mile south of the former. [SU 858-009]

KINGSTON (Old Duck)

Converted Remains of a White Smock-mill with a Fantail

In the front garden of 'Old Mill House' here at the top of Juggs Road stands a well-used bedstone. The working face carries the inscription E.S.N, 1899, and on the back, KINGSTON MILL. The date commemorates the building of a ground-floor 'house' by the Reverend Norris above the original stone and brick cellar base of Kingston smock-mill. This was popularly known as 'Old Duck'. Various reasons have been given for this, but I feel sure that the most likely explanation is simply that her unusual squat design suggested this fowl, especially when compared with the rather slim and elegant 'Shelly's' mill just across the Winterbourne Valley.

The large cellar base was built one storey high in an excavated pit and thus buried to give the required support to the rather thin walls. Two feet of brickwork was left exposed, and the wooden smock erected on the curb. Plate 132 shows the doorway into the base which is on the same level as the sunken Juggs Road and below the facsimile first-floor building.

Kingston Old Duck fell in 1891.[1] She had been in the hands of Mr. George Uridge, who was shown at Southover from 1843[2] till 1890,[6] and stopped some time earlier because she was found to be unsafe. It is said that the 'fall' was encouraged by the owner of Southern mill, who needed more wind for his mill, still working close by. Apparently, some of the brickwork was cut away beneath several cant posts and the whole 'thrown' with the aid of a cable.[3]

The first positive evidence of this particular mill occurred in 1802 when 'a substantial recently built smock mill'– 'leasehold at a ground rent of £1-5-0. Desirably situated in the Parish of Kingston'[4] was put up for auction. The following week's edition of the paper concerned,[4] carried the following fascinating advertisement:

> 17 December 1802. T. Pescodd & Son, Millers Southover respectfully inform their friends and the public, that their mill, lately erected is now ready for work. As no *toll* will be taken at their mill, if any deficiency should occur in weight on the meals' return from their mill (steamage accepted), they will give double the deficiency to those who shall apply. Pea, Barley and Oat meal for sale at their meal house in Southover on the lowest terms.

Presumably the Pescodd family were able to persuade clients of their honesty, for they worked the mill through good and bad times till 1828. She was then taken on by Isaac Leney until 1837,[1] when Mr. James Weller came. He was followed in 1843 by Mr. Uridge, as mentioned above.

Once a registered sea mark, Old Duck was a stout white mill with a broad hooded cap and a fantail.[5] Powered by four double-shuttered spring sweeps, she worked two pairs of stones underdrift, and spent her life in the company of Mr. Payne's Southern mill, of which, whilst outliving here, there is only a sunken swimming pool to mark the spot.

The site is to be found at the top of Juggs Road, which was once busy with traders who plied between Brighton and Lewes, and preferred this shorter Downland route. Today, the new Lewes bypass has cut the lane in two and must take the new service road which runs westward off the A275 to Newhaven out of Lewes. [TQ 402-092]

LEWES (Town Mill and Shelly's Mill)

Remains of a Large Bygone White Smock-mill with Fantail

The first photograph shows the converted base of 'Town Mill', built by Pilbeam of Lewes.[1] Along a quaint little alley known as Pipe Passage, this smock-mill base stands below the Keep of Lewes Castle in the centre of the town. The plaque tells us that the Town mill was built here in 1802 and re-erected on the hill to the west of the town about 33 years later (Plate 133), although chronological writings suggest a date nearer 1818.[2] Today we can still see the remains of the rebuild close to the north wall of Lewes Prison on the lower slopes of Race Hill. The illustration, *c.* 1870, is again one from the Reeves collection (Plate 135). Here, we see this fine example as it was when working as Shelly's mill. The Smart family had the mill from approximately 1812[1] until 1858[3] after which time Joseph Shelly took it on. Medhursts, the Lewes firm of millwrights, were involved in the move.[4] The smock-tower was re-erected on a single-storey brick base, and whilst the spot is open to all but the north, it was found necessary to raise the mill to improve its command of the wind after the prison was built.[2] Medhurst was employed once again in 1853 to raise the mill approximately 20ft. at a cost of £150.[4]

As can be seen in the second photograph, the mill worked with a large set of double-shuttered patent sweeps. These would undoubtedly have replaced spring or common sweeps originally fitted, as Cubitt's patent was not employed much before 1810. The cap which was an unusually shallow dome, had a copper sheathing and carried a fine Sussex fantail. A long steady pole for the weight chain hung down below this. Inside the white four-floor smock, the mill worked three pairs of stones, two pairs of French burrs and a large pair of peaks, plus dressing machinery.

The two men standing on the stage in the photograph would have been able to see Spital mill and Offham mill to the north, Kingston smock-mill and 'Southern' post-mill at work on Kingston Hill to the south, Malling post-mill to the east, and the six sweeps of Ashcombe mill to the west.

One William Hemsley was the last to work this fine seven-storey smock-mill in 1899, and Messrs. Wells, who took on Alfred Shaw's business, removed the sweeps and far in 1912.[1] Ten years later, the smock-tower was taken down, leaving the three-storey base. This was capped with a conical tiled roof. The third photograph shows this as it stands today, used as a tack store for Lewes stables.

The town site is best found by asking when near the Castle itself, as Pipe Passage is very close by and in a maze of lanes.

[TQ 413-101]

The Race Hill site is very close to the prison to the north of the A27.

[TQ 403-101]

LITTLEWORTH (Jolesfield Smock-mill) (West Grinstead)

Machinery and Base of Black Smock-mill with a Fantail (at separate sites)

Today all that remains of this mill is the brick base standing in the hamlet of Littleworth, and the machinery, which is shown here piled up on the front lawn of a restaurant (Plate 138). To find these we must go some 12 miles north of the original home to County Oak. There, at a site, which was in Surrey until April 1974, to the east of the A23, we can see the decaying remains of the driving gear.

This mill takes its name from Jolesfield Common as it then was. The site is a little to the north of this, though the common is not much in evidence today. The mill is shown on the first edition of the Ordnance Survey map 1813, and is first referred to in a notice of auction in 1788 at the *Green Man,* Jolesfield—'A new built accustomed corn windmill situated near the turnpike road from Steyning'.

Jolesfield is worthy of a fuller description, for it was an unusual mill in several respects. The smock-tower was larger in girth than most of the county's smock-mills (Plate 136). Above this, the cap rested on an iron live curb which was over 14ft. diameter at the rack. The curb had to bear the weight of the cap, the long iron windshaft (Fig. IV), the wooden clasp-arm brake wheel and the large patent sweep assembly. The cap itself was of an unusual beehive design with a large penthouse doorway extended from it to give access to the fanstage. The smock-mills at Washington (Plate 148), and Ockley in Surrey, had caps of a very similar design. Below, the smock-tower was large enough to accommodate four pairs of underdrift stones, two pairs of peaks and two pairs of burrs. This is unusual in any mill, especially in Sussex. That they were ever all used at once is doubtful, for all the available power from a strong wind is usually required for two pairs of stones and the dressing equipment in a mill of this size.

An old water-colour of Jolesfield mill by A. Foord Hughes,[2] shows her around 1930 looking rather squat and careworn. Whilst not having the functional lines of Chailey or Punnett's Town in its earlier days, both of Kentish line, she had charm and style of her own (Plate 136). Millers included W. Marshall, 1845; G. Rapley, 1866–1895; H. Nailard, 1895–1907; C. W. Pullen, 1907; and S. F. and H. Tidey, 1913–1927.[3] I believe that a Mr. Knight was the last to work this mill in 1928.[4]

The derelict mill was dismantled in 1959 by E. Hole for Mr. Hicmet, the owner, and removed to Gatwick Manor where, I understand, it was originally intended to re-erect the structure as an attraction. A curious building was put up instead, however, the the remaining machinery displayed at the mercy of the elements. This has suffered considerably since the photographs were taken in 1973. Here, in the grass, lies as a base the huge rack with its segmental iron rack. The winding worm lies nearby. Mounted on the ring are the main cap sheers, the 19in. octagonal oak upright shaft with its iron wallower and 8ft.-diameter wooden clasp-iron great spur wheel. Last, but not least, is the iron windshaft weighing almost two tons. Close to this 'montage' are two pairs of French burrs and their stone spindles. The brick base (Plate 137) is today a stable in the garden of a private residence.

The original site of the mill itself is two miles south of the A272 Cowfold/Billingshurst road in the village of Littleworth near West Grinstead.

[TQ 191-205]

The machinery is on the eastern side of the A23, two miles north of Crawley. [TQ 271-394]

LUNSFORD'S CROSS (Ninfield) (Thorne Mill)

Base of a Tall White Smock-mill with a Fantail

Reputed to be the tallest windmill structure erected in Sussex, with the exception, perhaps, of Shelly's mill at Lewes. 'Thorne' mill stood on rising ground to the east of Pevensey Levels. The site is not renowned for its windmills, and this one is first shown on the 1873 six-inch Ordnance Survey. The actual date of construction was not much before this and put at 1870.[1] A London firm of millwrights was responsible and were sued for £200 when the weatherboarding was later found to be faulty. The total cost of the mill is said to have been £1,200,[1] which in any case, even today, seems almost a gift compared to Shipley smock-mill which totalled £2,500 in 1879.

Thorne mill was indeed tall. A slender five-floor white smock stood on a two-storey tarred brick base. A Kentish hood carried a large six-blade open fan and patent sweeps. These worked three pairs of stones and modern machinery in the base. The mill could also be driven by a portable steam engine via an unusually high auxiliary power pulley.

Mr. Morgan worked the mill from 1870[1] till 1874,[2] when she was let and by 1882, Messrs. Simmons Bros. were tenants at £80 per annum. This family continued as millers here until 1905 when she became tail-winded. The result was rather spectacular, the cap hung, for a time, over the curb, countered only by the sweeps which had lodged against the smock (Plate 139). Apparently the mill had not been worked by wind for about a year before the accident. A photograph taken in 1906 shows that the sweeps were subsequently removed and the cap restored to its correct position. The body was taken down in 1907.

No one is recorded as working a business here from that time. Close to the base, which was eventually roofed in during the 1920s (Plate 140), is the remains of the granary and a little to the south-west stands the mill-house plus the bakery building. This still houses a brick baking oven. In the garden, between the base of the house, is a dome-shaped underground water reservoir which just shows above ground in the grass. This, built of brick, has a slate floor and once acted as a soak tank ingeniously supplied by a ditch dug around the mill base.

Found on the west of the A269 Ninfield to Bexhill road just before arriving at Lunsford's Cross. [TQ 714-109]

ORE (Black Mill)

Converted Square Base of a Black Smock-mill with a Fantail

Known as Ore Black mill, this one of a pair of smock-mills which stood in the village during the 19th and early 20th centuries.

First shown on the six-inch Ordnance Survey map of 1876, she was erected on a three-storey square brick base founded on sand stone rock between the years 1852 and 1855 for Mr. J. W. Thomas.[1] The similarity of this base to that of Staple Cross, next but one in order, is striking. Black mill was worked by the Thomas family for many years starting with John Wesley Thomas then by him in partnership with brother Joshua.[2] In 1869 she was advertised as comprising, '2 pairs of wheat stones with smut and flour dressing machines'. Mr. George Goldsmith took the mill on until 1874.[2] The registered tenant from then until 1882 was 'Horace Noakes[2]–Downs Mill'. The last miller was one Thomas Mills who was still here in 1890 (Plate 143), when the mill ceased working by wind, and the business transferred to Guestling smock-mill. The owner in the early part of this century was Mr. Raymond Hollands.[3] By this time, Black mill had stood idle for many years, and in spite of a short spell of work by auxiliary engine, she was pulled down at the end of the First World War.[3]

When last working by wind, this mill had a black wooden smock surmounted by a hooded cap which carried a white fan and four white spring sweeps (Plate 143). The three-storey tarred black base formed part of a stage for servicing these. Redundant and empty, this base was offered for sale at £100, and eventually purchased for £21.[3] The new owner, Rev. Arthur Tobitt,[4] utilised the building as a missionary hostel. This was sold once again in 1928 to Sir Henry New, who built the extensions that we see today.[4] Plate 144 shows the converted base which is paradoxically painted white over the tarring and forms the greater part of the house.

Found to the east of the A259 Hastings to Winchelsea road. [TQ 839-117]

ORE (White Mill) (Cheale's Mill)

Two-storey Brick Base of a White Smock-mill with a Fantail

A windmill is first shown here[6] at this site in the village of Ore on Greenwood's map of Sussex, 1823/24, and it was almost certainly the one which stood until 1900.[6] Unfortunately, Cheale's mill, as she was then known, was burned out in May[1] of that year. She was a white smock-mill, and in 1863[2] recorded as working three pairs of stones and equipped with spring sweeps (Plate 142). These were serviced from a stage which surrounded the mill above the two-storey base. The second illustration shows the remaining base today (Plate 141). This is in the garden of a house and part of a garage business here. As we can see, this is of brick and heavily tarred. I understand that this was roofed in and used as a store almost immediately after the fire. The business itself was transferred to Questling smock-mill,[2] which was still working at that time, half-a-mile or so north-east. She had already taken on the business of Ore Black mill described above.

The various millers recorded here were: Thomas Hembry, 1828;[3] Jesse Catt, 1858;[4] Streeter and Beck Ore mill, 1866;[5] William Draper and Son (who also worked Silverhill smock-mill in Hastings), 1874;[5] Isaiah Cheale, 1887-95; after which time, Feaist and Co.,[5] the last to work Ore White mill, took her on.

Easily seen on the east side of the B2093 as it descends to the T-junction with the A259 at Ore itself. [TQ 835-117]

STAPLECROSS (Ewhurst)

Converted Base of a Black Smock-mill with a Fantail

Here, close to the Sussex/Kent border, we find this converted base of a very well-known smock-mill. The first illustration (Plate 146) shows the mill in position perched on the two-storey square base similar to that remaining at Ore. The photograph here was taken approximately 69 years ago. This was a fine fantail smock-mill of the Kentish style. Today we can still see the mark of Thomas Martin, millwright,[1] on the front of the base: TM 1815 (Plate 145). Above, the support timbers for the verandah are still buried in the brickwork.

An item in the *Sussex Weekly Advertiser* confirms this when the mill was offered to let:[2] 'At Staple Cross, Ewhurst Sussex, a new erected stage mill with two pairs of stones and room for one more pair. For particulars enquire of Thomas Martin Proprietor', dated 11 November 1816. A Mr. Richardson worked the mill from 1855-1882[3] when mortgage problems came to a head with the Lewes bank. He was followed by Hilder and Son. Mr. Banister, the owner of several local mills,[3] had been proprietor for 17 years or so by the time it ceased work in 1916. Mr. Goble, the latter-day miller here, remembered the happenings that followed. The sweeps were left struck up when she stopped, and a few years later (1918) during a storm, she broke loose and began to run away. The villagers, quick to see impending disaster, helped to stop her by jamming poles and iron bars into the machinery, whilst others brought water in case of fire.[4] The owner took the precaution of removing the sweeps a little later. Mr. F. Grisbrook, a local builder, took on the premises in 1924.

When last working, the mill had a large Kentish hood which ran on a dead curb and supported a set of large one-and-a-half-shuttered patent sweeps, controlled by a modified Kentish lever mechanism.[5] These drove three pairs of overdrift stones within the three-storey wooden smock. Below, the two-storey brick base was used for dressing and storage purposes.

By 1951 the smock-tower was in such a dilapidated condition that the only safe way out was to demolish the remains. Mr. Dallaway of Punnett's Town smock-mill was present, for he required the windshaft for the re-build of his mill (see page 60). The electricity wires in front of the building were disconnected and the whole mill literally pulled off the top of the base. Unfortunately, the brake wheel was smashed, but the iron windshaft remained intact. Mr. Dallaway was able to take his prize and still remembers the celebrations that took place after. The rotating sweeps at Punnett's Town mill are a last tribute to this bygone.

The base, now a private house (Plate 145), is in the main street of the picturesque village of Staplecross. On the eastern side of the road which continues on to Bodiam, where a classic English moated castle is to be found amongst the hop gardens of a famous stout brewer. [TQ 782-225]

WARBLETON (Chapman's Town)

Remaining Base and Cut-down Tower of a Black Smock-mill with a Fantail

Chapman's Town is a small hamlet north of Warbleton. The mill here stood next to a ridge road which runs north to Punnett's Town, one-and-a-half miles away. The remain is rapidly decaying today (Plate 147), which is not altogether surprising, for it has stood in this form for 90 years or so. Neves, the East Sussex millwrights, still operating from Warbleton, were employed to dismantle the shattered mill after a devastating thunderstorm had raped the structure as colourfully re-told by a contemporary report[1] dated 20 July 1880:

> Warbleton
>
> Severe thunderstorms on Friday the 16th at Warbleton, the lightning being incessant from 5 am to 6.30 the thunder being most alarming. During the storm, the windmill occupied by Mrs Kingsland was struck. The electric fluid passed down one of the sails, entered the mill by the roundbeam, splitting it in three places and after doing considerable damage to the working gear, descended by the sack chain to the basement floor where several things were damaged and displaced in passing out. The mill door was shattered and a heifer standing by was killed—also some fowls.

Mrs. Kingsland must have found the whole thing rather 'shocking'. She was the last to work here, the mill being owned by Mr. H. Knight at the time.[2] Mr. John Chapman had been the previous miller and had, in fact, owned Chapman's Town smock-mill from its date of erection. This is rather illusive, for there is no map reference until 1873.[3] However, Mr. Chapman is shown here on the tithe map of 1838,[4] which places it sometime between Greenwood's Survey of 1823 and this tithe date. Mr. Chapman had the mill until 1876.

Unfortunately little is known of this mill's construction save that it was a black smock-mill with a hooded cap and a fan. The remaining two storeys of smock and sandstone base was roofed in with corrugated iron which was at one time painted red. Mr. Archie Dallaway has one or two pieces of machinery from this, including a steel yard from the tentering gear in his mill at Punnett's Town today.

I regret to add that the remains of the smock have now been removed.

One of the several smock-mill bases in this corner of East Sussex, this is to be found a quarter-of-a-mile north of Warbleton [TQ 616-184], not far from the site of Summer Hill post-mill, which only displays below ground pier footings to the persistent enthusiast now.

WASHINGTON (Rock Mill)

Converted Black Smock-mill with Fantail

Part of an attractive house (Plate 149), the home of the composer John Ireland between 1954 and 1962,[1] this is known as 'Rock Mill', The smock-tower is now covered with tarred felt and embellished with a white verandah just below the curb ring. The storm door still forms a recognisable part of a conical cap.

The picturesque site here is bounded by a steep-sided sand pit, which must be several hundred feet deep. This is part of a huge concrete and aggregate works which surrounds the mill ground on three sides. I first espied this mill many years ago from Chanctonbury Ring, which stands sentinel over this part of West Sussex. Whilst only a mile or so north of the Downs, the site here is open to the wind from all quarters, which drove this smock-mill perched on its mound of greensand.

Rock mill is understood to date from 1826/7[1] and is first referred to in a sale notice for 12 June 1837:[2]

> To be sold by auction on 14th July 1837 by Direction of the devicees in trust under the will of Mr Thomas Harwood deceased: A leasehold smock windmill on Rock Common in Washington with patent sails and all running gear therein. Capable of grinding 10 loads per week. Apply Mr. Henry Harwood Ashington.

Mr. Harwood was obviously successful in retaining the business for he was shown on the1840 tithe map[3] as miller. Mr. E. Mitchell followed, 1845-1855,[4] and by 1890 Mr. S. A. Coote was grinder for Mr. John Muddle.[5] At that time, this attractive smock-mill stood alone on the common. An old postcard shows her to have had a beehive cap with a penthouse extension similar to that of Littleworth smock-mill. This carried an eight-blade fan and a large set of one-and-a-half-shuttered patent sweeps. Mr. Muddle, the owner, who also lived at Ashington, was running her with two pairs of French burrs and one of peaks. I believe that the mill stopped work at the beginning of the First World War.[1]

The conversion here was carried out between 1918 and 1920 by a Mr. Redman who hailed from the West Country. At that time the mill still had sweeps up and machinery within (Plate 148). This was all removed and either used in the conversion, or sold for use elsewhere.

Easily seen as described above, but rather inaccessible otherwise.

[TQ 128-137]

WISBOROUGH GREEN (Champion's Mill)

Converted base of a White Smock-mill with a Fantail

Several dates of erection have been offered for Champion's mill, from 1805 to 1820, but it is not shown on maps dated 1813 and 1823/4. However, a notice dated 6 March 1820[1] describes 'an extraordinary good windmill situate close to the New River' and the 1842 tithe map[2] shows Mr. Thomas Welch here. The convoluted tributory of the Arun does, in fact, run within 200 yards of the site today.[3]

Here was a large white mill with a domed cap (Plate 150) which carried a fantail and spring sweeps. She worked three pairs of stones and dressing machinery and was described in 1860[4] as a Freehold mill in good working condition. The white smock-tower was mounted on an unusually tall two-storey sandstone base which also supported a stage for servicing the sweeps.

The Welch family were here until 1851 when Elijah Ford is shown, followed by G. Adams, 1862; A. Merricks, 1870; H. Ulidge, 1874; Elwyn Harrison, 1878; and H. W. Killick, 1899.[5] By the time that Mr. Killick was working here, the mill had also been equipped with two pairs of engine-driven stones in the base making five pairs in all. Trade ran slack around this time, and the windmill was allowed to become tail-winded in 1910, and, as at Lunsfords Cross smock-mill five years before, the sweeps jammed against the smock. Mr. Champion, the owner, made a number of spectacular attempts to right the cap with men and ropes which only resulted in a good number of well-meaning folk being hurled into the air.[6] He was, I am sure, able to restore their humour in his capacity as the landlord of the neaby inn. The cap and sweeps were eventually taken down six months later and the smock roofed in. Mr. Champion was not outdone, for I understand that this was used as local look-out point for several years. By 1915, the woodwork had become dangerous, however, and Mr. Baldwin, a local builder, was employed to pull the smock down and cover the base in. For this work he charged £20, less 5s. for firewood.[6] The sandstone base was used as a store for many years before becoming overgrown to the extent that, by 1960,[7] it was almost unrecognisable. In recent years this remain has been converted, very successfully, into part of a residence, and has gained a new short smock (Plate 151).

To be found in the centre of the attractive village of Wisborough Green, 200 yards down a short lane almost opposite the green itself and south of the A272. [TQ 050-258]

III

TOWER MILL REMAINS

ALFRISTON

Converted Cement-faced Tower-mill with Fantail

This mill remain is perched on the west side of the picturesque Cuckmere Valley just south of the ancient village of Alfriston. One of the true settlements of the South Saxons and steeped in history, this has a fine village square with its own market cross and surrounded by well-kept half-timbered buildings. In the summer this almost isolated spot seems to be a mecca for tourists from abroad, who, no doubt, have been acquainted with the 'Old English' spirit of the place.

Standing alone on the rising Down, rather sheltered from a good south-westerly, the mill here was built in 1834 on land then known as 'Rabbit bank'. This spot was leased to one Richard Saxby for 99 years at a ground rent of £5 per annum. During the following year the new tower-mill was put up for sale and subsequently worked until 1905, when it sustained a broken sweep at the hands, or should I say, horns of a cow. Millers[3] included Daniel Sudbury in 1845, William Shoesmith, 1855-66, followed by Thomas Harvey till 1874; then Thomas Fennell, 1881[1] when George Hewitt came.

A photograph in the possession of Mr. S. Ashdown shows the mill to have had a Sussex beehive cap with a fantail and spring sweeps (Plate 153). These worked a pair of peaks and a pair of burrs. Flour was still being produced at that time when a Mr. Hewitt was miller, but he ended up grinding for feedstuffs only. In 1907 Mr. Hewitt gave up the mill, and during the following year all the machinery was removed by Mr. Pearce from Dicker.[1]

Plate 152 shows Alfriston tower-mill today forming the eastern abutment of a residence. This was built in 1910 and has a modern extension. The cement-faced tower, like that of East Wittering mill, has an unusually steep batter, and may well have had an underdrift arrangement. Difficult to find, and on a private road west of that from Alfriston to Seaford.

Across the valley to the north-east is Windover Hill, and not far from the great barrow which crowns the site of the Long Man of Wilmington, is the site of Arlington (or Windover) post-mill. Remains of the roundhouse walls and a millstone may be found buried now beneath the turf. This little mill was burned down after becoming tail-winded 100 years or so ago, and from this lofty spot Alfriston's sheltered tower can be seen one-and-a-half miles away.

[TQ 518-027]

ANGMERING (Highdown New Mill)

Converted Once Black Tower-mill with Fantail

About two miles from the sea, this revived tower stands as part of a rather isolated residence (Plate 155). Originally called Highdown New Mill (Plate 154), it is sometimes known as Ecclesden mill, after the manor nearby of the same name. Until recently this was only the derelict remains of a fine tower-mill which had a beehive cap, a large set of patent sweeps, and a fine mill-house close by. This, working with two pairs of stones, was built for Mr. Grant in 1826[1] and took on the business of John Olliver's post-mill which stood nearby. Henry Grant is shown here for tithes in 1839,[2] he and his family working the mill during the whole of its working life, which ended in 1872.[1] In 1880 the cap and sweeps were blown off and their remains thrown into the precipitous chalk pit which forms the southern boundary here.[1] The whole had deteriorated to an empty, ivy-covered curiosity, by the time the Rev. P. Hemming came across it in the early 'thirties.

The ivy was removed some years ago,[2] and the brick tower, standing in three-quarters of an acre, is now part of a modern dwelling, as, I believe, a study, and above in what was the spout floor, a bedroom and bathroom. The whole was offered to prospective buyers in 1973 for £50,000.

Like most of the tower-mills in West Sussex, this one is situated on the coastal plain. This part of the county between the Downs and the sea, is very flat and is open to the prevailing wind coming straight off the English Channel. As we travel these parts, which we must to see the tower-mills at Arundel (a very similar mill), Barnham, Earnley, and Nyetimber, we will notice it is very productive. For here are hundreds of acres of glasshouses and nurseries bringing early vegetables and flowers to the south. The climate is kind here, and winter frosts very light when they occur, not to be likened to the hoary rime that covers the sweeps of the mill of the High Weald during mid-winter (see Nutley post-mill, Plate 52).

The remain is easily seen to the north of the road between Worthing and Littlehampton. Found to the east of the village of Angmering at the end of a long track which leads the walker to the local vantage point of High Down Hill. This recently-painted black tower makes a good landmark.

[TQ 082-044]

CROWBOROUGH (Pratt's Mill)

Converted Tiled Tower-mill with Fantail

Today we see that this is incorporated in a dwelling. Only three floors of the mill-tower remain here on the north-east side of Crowborough town, The site is somewhat sheltered from the prevailing wind. This, however, did not prevent the mill working for some 45 years by wind-power.

During late 1861 Richard Pratt, who became a Baptist minister as well as a miller, had the machinery brought from a demolished mill at Tunbridge Wells in Kent. This he had re-erected in his newly-built tower. In the book *The Wilderness Journey,*[1] Richard Pratt describes his calling to the Cloth and life at Crowborough. During February of the following year just after the mill was started a tragic accident occurred that must have taxed Mr. Pratt's faith, for his son, Jesse, was killed whilst working in the mill. The *Sussex Advertiser* carried the inquest notice and, dated 12 February 1862, reads:

> Richard Pratt, miller at the New Mill Crowborough, Rotherfield, deposed that the deceased was his son and was 18 years old. On Monday last (Feb 3rd) the deceased went into the mill to help him to dress some stones. He, the father, was called away after they had been stone dressing for about an hour. The deceased was working close to the upright shaft. The mill was going very slowly. Witness had just got indoors when William, another son, came screaming 'Jesse's caught in the mill round the upright shaft and I believe he is dead' (His frock was caught by the upright shaft and carried him round causing him to become entangled with the machinery). Samuel B Pratt, another brother, stopped the mill.

Mr. Pratt was not deterred and the family appear to have owned and worked the mill until 1922, latterly by steam.

The first photograph here shows the mill when still at work by wind, with two pairs of peaks and one of burrs, before the turn of the century (Plate 156). As we can see, the cap was a large hooded one and carried a Sussex fanstage, also a fine set of patent sweeps. These were sold in 1907 and the mill was subsequently worked by a producer gas engine.[3] I am indebted to the present owner, Mr. Holcombe, for his help in securing information and the photographs. The mill was bought by Mr. Holcombe's father in 1927 when the conversion was carried out. The second photograph shows this today (Plate 157).

Pratt's mill and Mark Cross tower-mill near at hand appear to have been very close kin. Both mills had a large Kentish hood and the slender tile-hung brick towers.

[TQ 518-311]

EAST WITTERING

Derelict Cement-faced Tower-mill with Fantail

Today this is little more than a burned-out tower. Although the mill here was not much more than a derelict four years ago, if we look more closely we find more (Plate 159 shows the mill as it was in August 1974). On the western side, hanging over the decayed breast beam is the windshaft, its canister long since empty. The sweeps which went about 83 years ago, were the well-tried combination of two 'springs' and two 'commons', favoured by many millers against all spring or patent sweeps. As can be seen, the cap and brake wheel have gone along with the fan. The cap roof was removed by a Channel storm in 1931. Here, as at Selsey, the mill is only a short walk from the sea and extremely exposed. The whole surrounding area is very flat, once covered by the sea, which after retreating, returned once again, taking in a monastery settlement south of here. The ghostly occupants can still be heard ringing the bells, according to legend. Substantial sea defences now safeguard the 'bill' hereabouts.

The cap sheers and truck wheels remained (Plate 159), although only just, as did the remnants of the fanstage. Just below, is the wooden curb which still has the majority of its morticed cogs in place.

The tower itself is an extremely sturdy brick memorial to master masonry of the past, being about 2ft. thick at the stone floor level with a rather steeply battered profile. This, I think, is because the mill is not unduly tall, and also that the stones, a pair of peaks and a pair of burrs, were under-driven.

At the time of my visit, I noted that some machinery and the floors were still in the mill, although collapsed to a great extent, as was seen through the windows. On finding this tower, I was reminded by its stocky appearance of the Martello towers which line this county's eastern shoreline from Rye to Eastbourne, erected some 200 years ago to repel our Common Market neighbours who were distinctly aggressive, Bonaparte being the chief offender of the day.

The first record of the mill itself is from 1810, when the owner was rated at two shillings per annum.[1] The first reference to a miller is that for Mr. Robert Woodman in 1845.[2] He was followed by William Stevens in 1855, T. Souch in 1866, R. H. Sparkes in 1870 (Plate 158), and Mr. E. Redman and Richard Stevens from 1887.[3] Mr. Redman was the last to work here in about 1895.[1] I believe that the sweeps were removed in 1896. The mill had suffered rather badly a number of years previous to this when the brake was applied rather suddenly. This resulted in a broken sweep and damaged the flour machine.[4] The fan was also a constant source of trouble. As mentioned above, the curb ring was fitted with wooden cogs which were meshed into the iron winding worm. These became worn and broken, allowing the fan to fun free and making it difficult to stop. Eventually this was arrested with a bale of straw. The miller then attached a long iron rod to a cap timber and pulled the whole into the wind by hand. The sweeps were at one time serviced by a travelling stage. I am indebted to the present owners, Mr. and Mrs. Redman for some of this information.

Today the mill yard is a garden, full of a family supply of vegetables, flowers, rabbits and pet guineapigs, and I have to add that the tower was completely burned out in May 1975. The windshaft has now fallen and lies within the base of the tower.

Not far from Selsey itself and easily found just to the north of East Wittering front on the road to Acre Street. [SZ 797-973]

EASTBOURNE (Hurst's or St. John's Mill)

Ground Floor of Cement-faced Tower-mill with Fantail

Approaching the town of Eastbourne from Polegate, along the A22, we find, when within two miles of the sea, that we are on high windswept ground. On the western side of this road a large 'mill house' marks the site of Ocklynge tower-mill which was blown up in 1936. Two hundred yards further down we encounter the *Hurst Arms* public house and a large church on opposite sides of the road. Turning to the east past the pub, we find the mill yard of Hurst's tower-mill. A large granary building still bears Mr. Hurst's advertisement panel, albeit rather indistinct today. A single brick storey of the tower remains as part of the business of Y. J. Lovell, Ltd., builders, the owners since 1941.[1] This is a traditional mill site and the tower-mill is understood to have been erected in 1808[2] to replace an earlier mill on ground which once belonged to St. John's Manor, Ocklynge. The Hurst family had worked the mill for almost a century when Alfred Stapley and Co. took on the business in 1903[3] as power-millers only.

Hurst's mill had four floors with a roundel extension around the base. The cement-faced tower carried a Sussex beehive cap, four spring sweeps, and a Sussex fantail (Plate 160). Two of the sweeps were sold to Mr. Ovenden for use at Polegate tower-mill after she finally stopped work some time in 1917.[2] The Rev. P. Hemming, in his *Windmills in Sussex, 1936,* has an extensive note of this mill for which he had a special affection.

When working, the interior machinery included three pairs of overdrift stones, a dresser, and an oat roller.

The mill had become rather derelict by 1936 and at that time stood bereft of sweeps and fan. The tower was cleared of machinery in the 1940s and reduced to one storey in 1950.[4] As may be seen (Plate 161), the roundel has also been removed. The whole area today is completely surrounded by modern development, and quite a contrast to the rural aspect enjoyed by Mr. Hurst 171 years ago. [TQ 599-001]

MARK CROSS (Walter's Mill)

Converted Tile-hung Tower-mill with a Fantail

This stands just to the left of the A267 in the village of Mark Cross, in a high position. From the top of the tower, Argos High mill could once be seen to the south, and Tunbridge Wells town some five miles north on the Sussex/Kent border.

Until the roofed-in tower was sold to Wealden farmers in 1925,[1] this had latterly been in the Walter family, handed down from father to son. It is interesting to note the lineage found in various directory references,[2] starting with Edward Joseph Walter in 1982, the story follows:

1887 Edward J. Walter and Sons.
1903 Edward and Joseph Walter.
1905 Joseph Walter & Son.
1913 Joseph Walter and Sons.

and then after the First World War:

Joseph Walter and Son once more.
Finally worked by Frank Walter.

The first reference to this mill that I have been able to find is for 1845[2] when one Catherine Ashby, farmer and miller, is recorded at Rotherfield. Similarly, in 1855,[2] Walter Dunk, miller and farmer, is shown at Mark Cross, Rotherfield. Mr. Dunk seems to have encountered difficulties and in 1856 the *Sussex Advertiser* carried the following notice:[3]

> To be sold by auction on premises under a deed of assignment for the benefit of auditors, all the live and dead stock and farm implements, stock in trade, furniture and other effects of Mr Walter Dunk, Miller & Farmer at Mark Cross, Rotherfield.

The five-floor brick tower was crowned with a Kentish hood, a fantail patent sweeps (Plate 162). She was reputed to have been the work of Arnold Brothers of Paddock Wood. Kent,[4] Three pairs of stones were fitted, two pairs of peaks and one of burrs. Latterly the mill was used for grist working only and had two pairs of 4ft. peaks, kibblers and cleaning machinery. These were worked by wind when possible and steam plant when becalmed.

On 26 July 1911[5] the mill suddenly caught fire in the middle of the day. The blaze was seen from Argos Hill and Cross in Hand mills. After this time, the tower was roofed over, given a castellated rim and then integrated with the adjoining building to form a power-mill proper. The whole was driven by a suction gas engine. Wealden farmers, Ltd., subsequently worked this and removed the battlements in the early 1930s. The tower was half tiled, rather like Polegate. The reason, of course, for tiling these extremely thick brick towers was to keep out damp which lodged for many months after a spell of wet weather. Heavy tarring and even metal cladding, as at Clayton, was often the answer to this. The tiling of these towers is always the most attractive but had a major drawback in that it is very difficult to secure tiles on a near vertical tower wall permanently. We find, however, that on Sussex mills they were laid in wet cement working from the bottom up, with the result that they came adrift over the years, as we find at Halnaker and Polegate.

Today the tower and power-mill are part of a dwelling (Plate 163), weather-boarding now covers the cement facing at the top, and copious window space added in the five floors.

The house is rather attractively set in a well-kept garden at the end of Mill Lane. A pity, perhaps, that, in common with other converted smock and brick towers, that a flat roof has been added, a cap or hood being more complimentary and attractive to the eye.

This remain stands at the end of Mill Lane, as described above.

[TQ 585-315]

THE NAVARINO MILLS (WORTHING)

Bygone Black Smock- and Tower-mills, both with a Fantail

I have included this note, as the Navarino mills were perhaps the finest of Worthing's mills and also because they were the last to stand in the town. The intriguing name of these mills first led me to delve further for information here.[1]

These are hardly mentioned in the wealth of writings on Sussex mills, although I cannot think why, for they were rather splendid and well-known in milling circles. The illustration here shows north and south Navarino mills around 1895 just after the north mill ceased work (Plate 164). The smock-mill (the further of the two) was the first to be built and is shown on the first edition of the Ordnance Survey map dated 1813. She was originally known as East mill, being the only one on this side of Worthing at the time. Two pairs of stones were driven by common sweeps which were carried by a Sussex beehive cap with a fantail. The sweeps were later altered to the much-favoured combination of two commons and two spring sweeps. The south mill was the work of a local millwright, James Sheppard, who built it between 1827 and 1829. A sale notice in the *Sussex Weekly Advertiser* is of interest here; dated 21 February 1831, it read: 'To let with immediate possession, a well built tower mill with self regulating sails, now in full trade, situate near the town of Worthing. For particulars apply to Mr. Sheppard–Millwright, Worthing'. This was equipped with a rather disproportionate fan and stage. The shallow ogee cap did nothing to enhance the mill's unusual and striking appearance. This also had two pairs of stones and could latterly be steam-driven via a pulley. The south mill was always called Navarino. This, I believe, commemorated the famous battle of Navarino Bay in 1827 when the combined British, French and Russian naval fleets not surprisingly, annihilated those of Egypt and Turkey.

The millers were recorded as follows:[1] North mill, Richard Hide till 1823, then H. Dale till 1838, followed by W. Streeter till 1845. South mill, Sheppard/Newland till 1838, then H. Dale till 1845, followed by Richard Markwick till 1851, and George Newland. The two mills were then taken on by Robert Newland who had a business in Worthing High Street. When Mr. Newland took on the south mill, the two became known as Newland's Navarino mills. The business was Newlands from 1866 until 1887, when it passed to W. Barker. The north mill ceased work in 1893/4, when she sustained damage to the fan,[2] and by 1900 the two mills were out of work and becoming derelict. I believe that the north mill was demolished in 1902, but the south mill lived on until, standing without fan and sweeps, this was also taken down to make way for housing in 1910[1] or so. These were the last of Worthing's many mills. The site is today completely built over and close to Navarino Road at the eastern side of Worthing front.

IV

HYBRID MILL REMAINS

REMAINS OF A HORIZONTAL MILL AT EASTBOURNE (Pashley Down)

The illustration (Plate 165) shows what is left to us today of a bolting house and a horizontal mill. I am indebted to Mr. Laurence Stevens for a full account of the occupants of this unique site.[1]

Here, high above Eastbourne on Pashley Down, is an old-established windmill site. A little to the north of this remain lies a depression in the chalk, this marking the site of a post-mill thought to date back to the 16th century.

Subsequently, Richard Budgen recorded a mill here in his map of Sussex dated 1724. A little later a bolting house was erected close by. This was a rectangular building with a mock Gothic façade,[2] overlooking the town and used in conjunction with the post-mill.

In 1752 Mr. Thomas Mortimer erected a horizontal mill on the site of the bolting house, actually incorporating the façade. Today we see remnants of flint walling approximately 50ft. in diameter and within these the excavated base of the bolting house itself. There were many variations of horizontal windmills, and the exact form of Mr. Mortimer's structure is lost to us. However, we may imagine this as a louvred smock erected on the flint base, containing a driving element similar to a large water-wheel mounted on a vertical shaft. The louvres were arranged to admit wind to one side of this rotor only. The power thus communicated was taken by the shaft into the building below, driving the stones in the usual manner. The success or failure of Mr. Mortimer's mill, which, incidentally, pre-dates Stephen Hooper's work, is not recorded, but he built at least one other in Eastbourne. A roundhouse which stood on the seafront until 1841 or so, was evidence of another of these. Erected in 1757, this was damaged by a storm and later converted into a dwelling. Prince Edward, who was Queen Victoria's father, visited here with his tutor during the 1780s.

Mr. Stevens is at present, with the assistance of his wife, preparing an archaeological report, to be published in the Sussex Archaeological Collections[1] in the near future, concerning a further horizontal windmill site on Ocklynge Hill. This, of similar size to Pashley, was built by Mr. Mortimer, *c.* 1767, and stood in some form until 1811, when it was burned down. The mill was apparently still being worked at that time, and by a member of the family, as 'Mortimers' Mill'.

Pashley Down horizontal mill worked for approximately 15 years, presumably in conjunction with the post-mill. By the 1780s both of these had become derelict, for a painting by S. H. Grimm dated 1785[4] shows the post-mill apparently with two sweeps and the base only of the horizontal mill. No mill was shown here on the first edition of the Ordnance Survey map dated 1813, but subsequent maps by William Figg, 1816 and 1817, show this as 'Horizontal Mill in ruins'.[3] Archaeological evidence suggests that both mills decayed into ruins and that the whole site was extensively pilfered.

Laurence Stevens, Richard Gilbert and others undertook an archaeological survey of the Pashley Down area in 1966. The post-mill site revealed a circular clunch wall at the base of which were traces of trestle supports and on clearing the bolting house walls, several contemporary items were unearthed. Work at Ocklynge Hill, one mile north, in 1970 and since by Mr. and Mrs. Stevens has revealed possible sites of two medieval sunk post-mills as well as the further horizontal mill.

This remain is easily found by walking either from Paradise Woods up the hill or from one of the roads that pass nearby.

From this part of the Down, overlooking the sea, the town of Eastbourne is seen spread out to the east, stretching away 500ft. below us, towards Pevensey Bay. The mills hold the honour of marking the most easterly flanks of the South Downs which end abruptly here, and at Beachy Head to the seaward. [TV 593-932]

REMAINS OF A UTILITY MILL STRUCTURE AT DITCHLING (Elphick's Farm)

A square brick base is all that stands here today (Plate 166). The mill was erected by Mr. Richard Oxley Hollis here at Elphick's Farm in about 1884.[1] The structure originally carried common sweeps. These were changed to patents when the whole was re-worked some little time after it was built.

Mr. F. W. Gregory, who has been kind enough to supply the photograph here (Plate 167) knew Mr. Hollis's son and was able to record and supply me with details of this true utility structure. The drive was taken from the short windshaft by bevel gearing and transmitted down through the central post to the machinery floor in the base. The whole thing appears to have been a home-made arrangement and a product of unusual ingenuity. The upper part of the drive was quite open and supported by a rotating tree trunk, a branch and additional strut formed a fork which carried the two windshaft bearings. The upright shaft passed through, and was supported by, this post. Winding was achieved by pulling the whole into the wind by means of a tailpole or lever within the base. A trestle structure incorporated in the roof timbers steadied the post forming a wear or bearing collar at the apex and a pintle support in the cross trees. This, I feel, can only be classified as an unusual, if not unique, variation of the hollow post-mill and certainly on its own in the latter days of Sussex windmilling.

Wind power was used for driving a chaff cutter or saw from a line shaft by bevel gearing and a small pair of stones, also by bevel gears. It is likely that the power derived by the small sweeps would only have been sufficient to work one of these pieces of equipment at a time. Mr. Hollis must have been successful here, for he worked by wind for 16 years or so until 1900, when an engine took on the work. The sweeps were taken off in 1904.[1]

Today the farm is a private residence and most of the outbuildings converted into stabling.

Easily seen from the B2212 Haywards Heath to Ditchling road a mile north of the latter. [TQ 322-169]

CHAPTER SEVEN

PEOPLE

THERE IS ONE FACT that has always impressed me about the people associated with Sussex mills and milling, and I am sure that this is common to other counties. When visiting the many mills and sites, I have found a genuine friendliness towards those with real interest in the subject. These people connected with this industry of an age gone are real Sussex, and are as much a part of the scene as the countryside that surrounds the mills.

MR. S. ASHDOWN–MILLER AT CROSS IN HAND

Mr. Sidney Ashdown is the last in a long line of family millers. Great-great-uncle Eli worked here at Cross in Hand between 1850 and 1862 (see page 145), some years before grandfather Jabez came here as grinder for Mrs. Kenward in 1882. (Six years later, Jabez took on the mill in partnership with Mr. J. B. Newnham.) His father, John, continued from 1926 until his death in 1936. Sidney then took over at the age of 18 with his father's example to follow.

I met Mr. Ashdown at his work close to the windmill he had worked in for nearly 40 years. The business of Newnham and Son was as millers and corn merchants (see Chapter Five), and worked with feedstuffs such as barley, oats, maize, and beans, occasionally grinding wheat for special orders of wholemeal flour, using the windmill and power-mill here. Mr. Ashdown (Plate 170) told me of the ease with which the power-mill could be worked, compared with the windmill. Setting the mill stone 'tenter' to give the meal the correct texture, keeping the feed rate constant, ensuring that the sweeps were set to take best advantage of the fickle wind are the various tasks that the stable environment of the power-mill makes light of. Vertical mounting of the stones made their support and tentering more reliable. The grain feeds in the windmill are controlled from the spout floor by cord lines which hold the feed shoes at the required angle above the eye of the stones. Moisture and temperature caused slow changes in the settings; the miller would check the texture of the meal regularly and perhaps give the twist-peg on the feed-line a quarter turn, then nip it tight again when he was satisfied. The power-mill here had mechanical feeding which once set against the constant stone speed, did not need further attention from the miller until the work was done, except perhaps a drop of vegetable oil in the various bearings.

Still evident in Mr. Ashdown's case is the agility and quickness of hand and eye coupled with a neat efficiency that are needed by a good miller in a working windmill. There were ladders to climb to raise sacks by the hoist operated in the bin floor. He had to stop the mill to bring in the second pair of stones or the kibbler if the wind was right, then go down to the spout floor to check the tenter and feed settings. Moving the filled sacks of meal about and raising sacks of grain into the mill for hoisting through the trap doors to the bin floor was no light task either. Then, perhaps, the weight box needed attention as the wind eased or freshened, indicating a change in work-load.

When not at work, there was always the dressing of stones (Plate 170), especially if, as between the wars, 'Sussex ground oats' was produced, the stones needed attention inside a fortnight. Mr. Ashdown still dresses stones for other mills, both wind and water, preserved in the county today. This is a country craft which is fast dying. The skill and accuracy with which the cutting edges are sharpened after very careful levelling, using a wheatstaff and red ochre, is difficult to match today. Sitting on the stone, the stone dresser chiselled, cracked and stitched with hardened steel bills and picks for many hours to give a perfect nip and cut to the grinding operation.

There are a number of stones close to Sussex mills for us to inspect, including those in the Mill Museum at Polegate. We can see the intricacy of the cutting edges and 'stitching' across these stones' surfaces.

The qualities and skills mentioned above are still very much with Mr. Ashdown in company with a traditional knowledge of mills and milling in Sussex, which is much enhanced by his unrivalled collection of photographs both old and new. This book owes much to evenings spent with Mr. and Mrs. Ashdown, talking of his mill work and looking over the photographs. He and his wife often go over to Burwash to Park water-mill. This produces wholemeal for the many visitors who come to Batemans, the one-time home of the Sussex author and poet, Rudyard Kipling. Sid is honorary miller and is pleased to exercise his skill both as miller and stone dresser, which, fortunately for future generations, is not lost. This water-mill has been painstakingly restored for the National Trust and works a two-hour stint when the water level permits. (Sid is shown dressing the peaks at Cross in Hand mill in 1962; here he is using a bill to re-work fine grooving into the furrows after stitching the lands.) A visit to Sid's home revealed all the dressing tools. Plates 171 and 172 show a selection which includes various bills, picks and thrifts. The rubbing stone illustrated is used for an initial smoothing of the lands; this is carried out prior to proving with the wheat staff and red oxide. The spectacles shown are to protect the eyes from flying particles of stone and fragments of hardened steel from the tools. One of Sid's mill bills, made by Chambers of Canterbury, was given to him by an old friend and mentor, the late John Russell of Cranbrook. (Mr. Rex Wailes in his book, *The English Windmill,* has included an account of stone-dressing technique by Mr. Russell).

Of the future, Sidney Ashdown hopes to see the revival of interest in the county's wind- and water-mills in general, continuing to thrive. In particular I feel sure, the likelihood of his family mill working once more is very dear. This was a fine post-mill and one which Samuel Medhurst must have been proud to have been associated. It is to be hoped Sid will soon be miller at Cross in Hand once again.

MR. G. A. CATT OF HAMLIN'S MILL, HAILSHAM

As previously mentioned under the heading of this bygone smock-mill, Mr. Gilbert Catt, still active at 73, lives close to the site of the family mill. With a lifetime of experience of and enthusiam for Sussex milling, Mr. Catt recalls working in the mill as a young man as his father, Ebenezer, and grandfather, David, had before him. At the time, he used to work with two of the three pairs of wind-driven stones, one pair of burrs and one pair of peaks. (Only two pairs were used latterly for provender production.) The business latterly baked bread from proprietary white flour which was bought mainly from the Britannia mills at Portslade. The output of the windmill, which supplied feedstuffs to the local poultry-fattening industry and ground farmers' corn for their own use, was capable of an average of five tons a week at this. When grinding oats, the output was rather less. Mr. Catt worked with oats, barley, beans and maize, and described the processing as follows:

1. Oats were either rough ground through peaks for cattle feed or put through well dressed and tentered peaks to produce Sussex ground oats.

2. Barley was ground whole between well dressed peak stones for meal.

3. Beans were always kibbled before grinding between peaks to produce a rough meal.

4. Maize could be ground whole between a pair of French burrs which could cope. Alternatively, it was kibbled and then put through peaks. Both methods produced 'fine maize meal'.

The production of Sussex-ground oats was particularly demanding, and the peak stones were dressed after something over a full week's work at this. Stone dressing was another of Mr. Catt's 'apprentice' tasks and for fine working with peaks, he would employ finely-stitched lands and close-grooved furrows. A pair of emery composition stones similarly used in the mill later only required a trimming out of the furrows every fortnight or so.

By the early 1920s Gilbert Catt was an experienced miller and continuing the family tradition, but work in the windmill was to come to an abrupt end. One night in November 1923, Mr. Catt's grandmother awoke the household with the news that the mill was on fire. Young G.A. was sent to Hailsham (on foot) to raise the fire brigade, and remembers seeing the maroon go up as he approached the town, for an alert neighbour had beaten him to it. Mr. Catt turned around and could see that 'all was lost'. The mill, which had been clad in iron some years earlier, was burning like a huge chimney ablaze, the weather proofing containing and intensifying the inferno which had started in the base. Nothing could be done but watch the tragic end. Later in the night, the sweeps and windshaft fell out, harmlessly, into the adjoining field, and by daylight there was little more than a tangled mound of iron and wet grain amid smouldering timbers.

The business was re-established as a power-mill and bakery and continued until 1967 when Mr. Catt retired. Like his fellow miller, Sidney Ashdown, his work has become his hobby and he is now miller at Michelham Priory. A long and expensive restoration has recently been completed. I found Mr. Catt here—grinding for wheatmeal between French burrs which are powered by newly-installed driving-gear, a great deal of which is reproduction. Mr. Catt is shown here checking the sackhoist (Plate 173).

MR. ARCHIE DALLAWAY OF PUNNETT'S TOWN CHERRY CLACK MILL

Having mentioned several features of Sussex mills that are unique to the county, here is a man who is perhaps unique in milling in the country (Plate 174), for he, in his time, has had to be millwright, wheelwright, carpenter, sawyer, engineer, and miller. Evidence of this is easily found in his smock-mill (see Chapter Five). Once again we find a grandson still in command. Mr. Archie Dallaway's great-grandfather, Samuel, had the mill brought here in 1859 from Kent. The family business, which was variously as millers and sawyers, flourished and waned until the late 1920s, after which time the mill here was dismantled and left an empty shell. A small power-mill worked for some time after. Since 1946 Mr. Dallaway has devoted his time to the reconstruction of the mill. Often working from first principles, Archie has, as far as possible, kept faithfully to milling practise. He has had to search out contemporary machinery for his mill, the components coming from a number of demolished mills. The iron windshaft was brought from Staple Cross smock-mill, for example. A few improvised and cleverly camouflaged modern equivalent components are used. For example, the governor weights for one of the pairs of stones are, in fact, oil filter bodies taken from tractor engines and filled with the required amount of lead. The crutch poles (overdrift driving spindles for the stones) are fabricated from steam tubing and motor universal coupling claws. True preservationists will no doubt frown at this sort of improvisation, but we should perhaps reflect that the industrial revolution in England started in remote areas like the Weald of Sussex, and it was just this sort of country engineering that helped to bring British technology to its latterday peak. Mr. Dallaway is, in fact, a true country engineer. With a broadness of outlook and skill, nearly all rule-of-thumb handed down from his great-grandfather's day, he has also been able to use in the mill timbers from personally selected trees, using remarkable thrift in an otherwise extremely expensive exercise. We can see the expertly-fashioned wooden brake wheel and true clasp-arm great spur wheel. Every tooth in each of these wheels was personally fashioned and tested with great care. This is a remarkable achievement on its own.

A certain amount of opposition to the mill's existence was expressed at one time, but with the help of friends, where extra labour was needed, Mr. Dallaway has been able to win the admiration of 'some people' as he put it.

Mr. Dallaway is now 68 and we must hope that his pride and joy will remain as it is today, a tribute to a memorable talent, for a very long time to come.

MR. FRANK GREGORY–A UNIQUE SUSSEX WINDMILLIAN

Frank Gregory has had the well-being of Sussex windmills and water-mills close at heart for close on 45 years. Starting as a youthful enthusiasm, this unique association has become a way of life. Many windmills in the county have had the benefit of his attention both in an advisory capacity and as a result of restoration efforts by him.

Anyone who knew the mills at Polegate and Nutley in the late 1950s will, I am sure, admit a debt of gratitude to the preservation groups that have brought these from near dereliction to the restored state to be seen today. At Polegate, Mr. Gregory played an important part in the organisation of the programme which, involving work both amateur and professional, resulted in a complete restoration of the whole mill. This included a new fanstage, a set of working sweeps and refurbishing the interior. In the adjacent museum, there are photographs taken as the work progressed and a number showing the state of Sussex windmills through recent years. A fine working model of a Sussex post-mill takes pride of place at one end of the room. This, exact in scale and detail, is the result of hundreds of hours spent by Frank with the tools of his profession.

A full-size example of this dedication can be seen at Nutley post-mill. The work here, which recently gained an Architectural Heritage award, was almost entirely carried out by the Nutley Preservation Society. The group included Frank Gregory, Ken Hamlin, Tony Shaw, and Brian Pike. Apart from organisation work, Frank's practical involvement was in setting out and fitting some of the main timbers of the body and the trestle. One particularly gruelling task served to give new life to the ageing crown tree which had, for the first time since 1928, to take the greater part of the weight of the body. It was decided that a steel joist should be inserted in the top of this beam. A groove was cut 8ins. deep and 3ins. wide and 10ft. long, an arduous task, with a broad chisel in oak. The group no doubt felt well satisfied when the supports were finally taken from the underside of the main spout floor sheers.

Photography has been a major part of the painstaking record work which any dedicated enthusiast must involve himself. This has served to put on permanent record the state of the county's mills, both wind and water, throughout the period of Mr. Gregory's involvement. Another aspect of this side of his has required a personal contact with the millers and owners. From these, particularly the friendship of the late Peter Powell, the last miller at Shipley smock-mill, and Sidney Ashdown, the last miller of Cross in Hand post-mill, a working knowledge of Sussex mills has been gained. This first-hand experience is hard won today and counted as a rare privilege.

When the mills at Shipley and Nutley are open, visitors are likely to be guided over by this windmillian and if fortunate, may even see a little corn being ground into wheatmeal. This Frank endeavours to do at least once a year, wind permitting. It will be remembered as mentioned earlier that Shipley mill is a large one and complete control of her requires the right amount of strength, knack and confidence, especially on a blustery day with two dozen enthusiastic visitors attending. I remember arriving on a particular Sunday afternoon, to find that she was tail-winded against a very light breeze. Was there any hope of grinding, I wondered?; could she be hand-wound to bring her sweeps into the wind, for what it might have been worth? We went up to the fanstage; Frank disengaged the fan, and with the hint of a grin, handed me the winding handle. After a few very strenuous turns, it was obvious then any miller here must have cursed the dead curb under the cap on many occasions. The view over the mill field and across the Low Weald to Chanctonbury Ring was, however, some consolation.

The academic side of things, so to speak, requires visits to the Society for the Protection of Ancient Buildings' meetings in London, attendance at committee meetings at the Sussex Industrial Archaeology Society, and at regular gatherings of preservation groups past and present.

Apart from the commitment to wind- and water-mills of the county, Frank Gregory is a countryman and takes great pleasure from walking. Many Sussex ramblers owe much to years of route planning and organisation by him.

Holidays, not surprisingly, often take him walking or to other mills here and abroad. There is usually one close by or on the way there, he told me.

Although now retired from work as a senior woodwork master, his love for wind and water-mills still takes him all over the country, and he is very much involved in current restoration work. At Clayton the recently-formed Jack and Jill Preservation Society, which includes a group of Sussex Industrial Archaeology Society members, is to bring Jill post-mill to working order with the combined use of professional and amateur labour (Plate 175). Frank's experience and guidance is invaluable and one of the project's greatest assets.

MR. R. HALL OF STONE CROSS

One of the first people I encountered on my exploration of the mills of Sussex was Ron Hall of Stone Cross (Plate 113). Mr. Hall, who hails from Dicker, bought the tower-mill at Stone Cross from the last Mr. Dallaway to work the mill here (see Chapter Five). In 1966 when Mr. and Mrs. Hall came, the mill was rather neglected. Since then Mr. Hall has taken on the task of renovating the mill, and has had a long toil with local authorities and preservation-minded people, without much financial help. Mr. Hall is very proud of his mill, and has an enormous job on hand.

Seven years ago, I found the mill surrounded by scaffolding. The fanstage was in the throes of renewal. Mr. Hall has since removed the rotting timbers and completely rebuilt this. There seems to be more than enough work for one man. Let us hope that he is able to carry out all that he would like to, for his object is to bring the mill to a working condition with a small museum and a perhaps a small zoo. May all luck be with him.

MR. C. HOAD OF BEXHILL DOWN MILL

Here in his office close to the remaining roundhouse and trestle of the post-mill (see page 95), I found Mr. Clifford Hoad, the grandson of the Mr. James Hoad who came here in 1857 as grinder to Mr. Reeves. He still heads the family bakery business.

The late-19th century business was in grinding and concerned with converting wheat into flour, also barley and oats into feeds for the local farmers. The womenfolk of the family began to find that they could easily sell to the farmers bread baked in a brick oven at the house here. The farmers would enquire, 'Is your dame baking tomorrow?' As time passed, the bakery proper was established, and both the wind- and power-mill business passed to the present Mr. Hoad's father, John. During the First World War, due to the blockade of Canadian grain boats by the Germans, coarse flour was used more and more for baking, although this was often of very poor quality, being part maize and beans. The windmill was used in conjunction with the electric mill at this time for grist work and after the war a small amount of flour grinding as well. The present Mr. Hoad was literally given the business after his father, John Hoad, had bought out the other sons and daughters. Here, since 1928, when he married, Mr. Hoad and his wife have raised a family of five girls and two boys. Today the family are still involved in the running of the business.

Bread and confectionery are produced in the bakehouse close by from proprietary brands of white and wholemeal flour. A shop on the site is also run by the family to sell some of these. The rest is delivered over a wide local area.

Mr. Hoad is over 71 now and, like Mr. Dallaway of Punnett's Town, and Mr. Ashdown of Cross in Hand, is an established family miller. Unlike his contemporaries, however, the family association seems to continue on. This fact must please him, and in itself is rare if not unique in Sussex today.

The scene which Mr. Hoad oversees has changed little since his father's day, and whenever visited is full of the chatter of cheerful people. There is something here for all of the family, including a farmyard quota of birds and animals. Chickens, ducks and geese, which, like a family of rabbits, are free range, living at peace with a dog and cats.

Of the past, Mr. Hoad remembers the working of the windmill well, and as a young man had charge of the grinding and baking. The photograph here shows Mr. Hoad with the thrift and mill bill last used for stone dressing (Plate 176).

Grist working gradually ceased when large mills started to market so-called balanced feeds with special protein and vitamin additions to patent formulae usually in pellet form. Mr. Hoad had his own feed formulae for domestic animals using barley, oats and maize with crushed shell added as grit for chicken, plus cod liver oil for pig nuts.

It would be sad if this portion of the Sussex past is lost to us in the name of progress.

MR. C. J. NEWNHAM OF CROSS IN HAND

At 74 years, Mr. John Newnham is a remarkably active man, involved a great deal with his work on the County Council (Plate 177). The business of grinding commercially having ceased in 1971, Newnham and Son became part of Bibby Feeds. Mr. Newnham has a wealth of information on his family association with Sussex milling, both wind and water. His great-great-grandfather was one of the early millers here at Cross in Hand itself, working the old post-mill between 1777 and 1783. Mr. Newnham was kind enough to show me a little book called *Memoir of Ali Ashdown.* This was written by a well-known character in mid-19th-century Sussex windmilling. Eli was Mr. Newnham's maternal grandfather and worked the old post-mill as apprentice grinder, loader and ploughman in 1850 for Mr. Hallet at 6s. a week. Three years later he went to work for Mr. William Kenward at Mount Ephraim for 10s. a week, and came to Cross in Hand once more in 1855 with the new mill. Initially he worked the old mill which Mr. Kenward had taken on in 1854, and tells us that in the summer of that year he assisted the millwright in still weather to put up the new mill. As grinder, his work was conscientious and, like Richard Pratt of Crowborough mill, he had a calling as a lay preacher. Mr. Kenward sent word to him one Sunday morning in 1862 to work in the mill as the wind was good. 'I have worked the night through and will work tonight through till tomorrow, but will not work on Sunday'. Mr. Kenward's reply was short. 'Tell Eli that if he will not work for me on Sunday, he'll not work for me on a weekday either'. Eli was not outdone, however, for he was soon to prosper as miller in his own right. Working 'Southern' post-mill at Kingston in 1866 with a partner, he brought his business to 22s. 6d. per week in a short time!

Mr. Newnham's paternal grandfather worked several water-mills locally, and finally took to farming. He prospered well enough to head a large family in comfort. Here the business link between the Newnhams and the Ashdowns was established. The family link was already close as Mr. Newnham's parents were both cousins to Mr. Ashdown's grandparents, so it was in 1888 that Mr. C. J. Newnham's father John, and Sidney Ashdown's grandfather, Jabez, became partners in business, initially working the two post-mills at Cross in Hand. This ran for over 80 years, each generation continuing the work of the last. For many years now, Mr. Newnham has been trying to raise the necessary funds to bring his mill to working trim once more. Quite a lot of repair work has to be done at this time, although the mill is still fully equipped to work. A new pair of steel stocks are waiting to be hung, and perhaps in the near future we shall be able to enjoy the successful conclusion of his efforts.

JOHN OLLIVER (THE MILLER'S TOMB) (WORTHING)

No survey of the Sussex milling scene would be complete without reference to the last resting-place of Mr. John Olliver, known locally as the miller's tomb (Plate 178).

About three miles to the west of Worthing is the local beauty spot known as High-Down Hill. Although not of great altitude, comparatively speaking, this hill-fort site rewards the walker with a wonderful view over the coastal plain, the sea and the southern slopes of the Downs. Unlike the similar aspect of Halnaker Hill to our west, there is no mill here today. However, if the walker has come here from the eastern side, he cannot have failed to notice the impressive tomb which stands as it has done for 210 years or more on the windswept approach.

John Olliver, who had previously worked the post-mill on Lancing Down,[1] came to High Down Hill *c.* 1750. He took on a post-mill which stood on the south-western corner of the camp.[2] The mill remained until 1826/7, when his family sold the machinery after having the structure pulled down.[3] When I last visited the site, the ground had been cleared of scrub and it exhibited a circular patch, typical of that swept by a tailpole, and a central depression. In 1766, when Mr. Olliver was 57, he made preparation for his death by building his own tomb. It is fair to say that he was a morbid character and is reputed to have spent his latter years awaiting his own end. His funeral gives us an indication of his outlook on life. The tomb, of course, is on unconsecrated ground, and the burial service was read by a young girl. This included a short lesson indicating his distrust of every man and yet acceptance of his Maker's final judgement in all things. Two thousand mourners were present to witness the burial and were dressed, as dictated by the miller, in cheerful clothes. The event took place in 1793, and John Olliver had lived for 84 years!.

The location of the tomb is unmarked on the present Ordnance Survey map, but is worth hunting down. To be found a quarter-of-a-mile north of the A2032 Worthing to Littlehampton road. [TQ 095-042]

CHAPTER EIGHT

SUSSEX MILLWRIGHTS

DURING THE EARLY PART of the 19th century, the windmill population of the county reached a peak, and of the mills which stand today, over half were built during the first 30 years or so. In the 1820s there were approximately 20 professional millwrights at work, and no doubt were making good livings. However, by the 1870s this total had dwindled to half-a-dozen or so traditional millwrights.

When considering the latterday windmills of the county, we find that the famous firm of Medhurst was associated with post-mills, especially in East Sussex. Neves, that is S. Neve, then Neve Brothers, were involved with work on the smock-mills in the east, although Hubbard and Upfields were associated with earlier examples.

Neal and Cooper of Henfield seem to have covered post-, smock- and tower-mills during a long period of work in the Mid and West Sussex areas. Holloways of Shoreham began to thrive, especially concerning the tower-mills in the west, as Coopers declined. Here again, these two followed earlier millwrights such as Blackaller of Steyning, Urrey of Chichester, and Sheppard of Worthing and Arundel. One or two isolated examples of lesser-known millwrights' work are still to be seen. The attractive Victorian smock-mill at Shipley is an outstanding case in point, the work of an appropriately-named Horsham firm, Grist and Steele. It must be said here that of all the mills covered in this book only a handful can be positively identified as the work of a particular millwright. However, I think that it is fair to say that the majority were worked on to a greater or lesser degree by one or other of those mentioned below, namely: Cooper of Henfield, Holloway of Shoreham, Medhurst of Lewes, and Neves of Heathfield.

Unfortunately, none of these survive today, but Sussex does have one modern-day millwright, and whilst not traditional, Hole and Son of Burgess Hill have built up a reputation in this field during the last 20 years and deserve a place here.

All the traditional millwrights were concerned with the broad spectrum of their trade which of course included water-milling. In the case of Medhurst, whose name is to be found in many wind- and water-mills today, we can see in the photograph, the water-wheel of Fletching mill. There were, of course, many firms of engineers who, in their time, were involved in millwrighting to a small degree, but their evidence is less easy to find (see Index list).

We can, however, find many names of suppliers to the trade generally on ancillary parts, such as the stones themselves, dressing tools, flour machines, and alike. Famous for many years were such as W. R. Dell, Armfield, Artingstall, Corcoran, and Tattersalls. Power-driven vertical stone-grinders were made by Barren and Turners. Old windmillers must have regretted the rise of others like Christy and Norris, Porter and Simon, for these firms manufactured the roller- and hammer-mills which work with wheat and provender in modern power-mills.

COOPER OF HENFIELD

James Neal, the founder of this firm, worked from Henfield and is first shown in a directory for 1839[1] and again at his property in 'Nep' town for tithes in 1844.[2] One of the earliest windmills on which he worked stood not far from the village centre and to the south of the common. The inscription 'J. NEAL MILLWRIGHT 1844' was to be found on the ironwork within. This was reworked as a spur gear post-mill with tailpole fan-tackle. Sadly, after suffering various mutilations such as an observatory in the roof, this post was pulled down in 1953.

By the middle of the last century this firm had become Neal and Cooper (the pentrough of Woods mill bears the cast inscription 'NEAL & COOPER MILLWRIGHTS 1854') and from this time they seem to have worked on many windmills and water-mills in Mid and West Sussex. Examples of their particularly well-finished ironwork was to be found in a number of them at one time, and the period of maximum activity appears to have been from 1860 onwards, and *Kelly's Directories* show that this was as William Cooper, millwright, until his death in 1876. At Ashurst, not far from Henfield, in 1864, they re-fitted the post-mill, and, I believe, were responsible for fitting an iron windshaft and brake wheel and patent sweeps. West Chiltington smock-mill also contained ironwork fitted by these local millwrights.

The most advanced example attributed to them was, unfortunately, completely gutted in 1912 or so; this was the large tower-mill built at Clayton in 1866. Of the ironwork left which includes the windshaft, the supports for the staging are very ornate. Holloway, the millwright from Shoreham was associated with Jack, but not until later, I think. Cooper's work on water-mills was evident, and their mark (W COOPER MILLWRIGHT & ENGINEER HENFIELD 1871) is still to be found on the pentrough of Hammond's

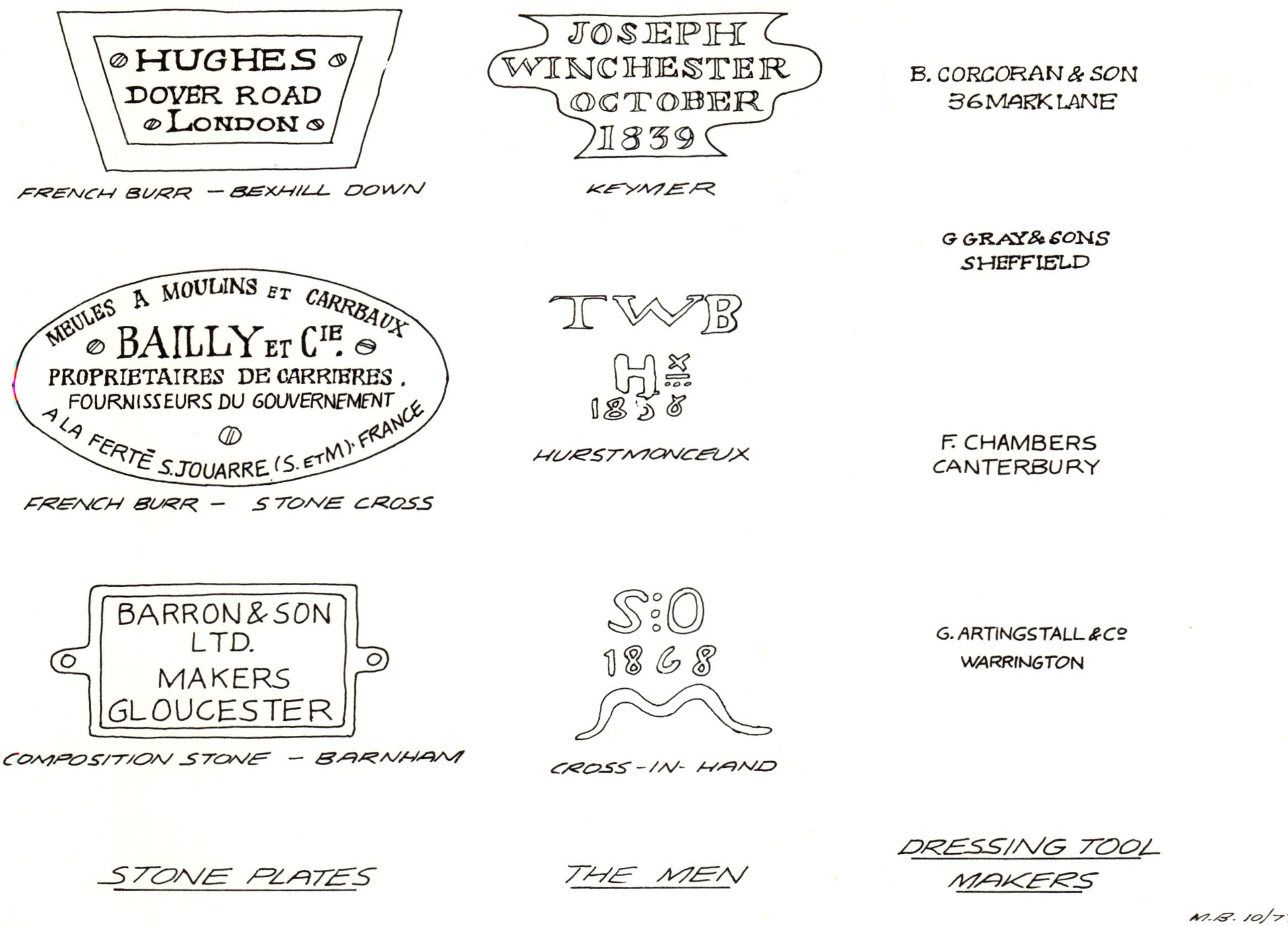
HUGHES
DOVER ROAD
LONDON
FRENCH BURR — BEXHILL DOWN
JOSEPH
WINCHESTER
OCTOBER
1839
KEYMER
B. CORCORAN & SON
36 MARK LANE
G GRAY & SONS
SHEFFIELD
MEULES A MOULINS ET CARRBAUX
BAILLY ET CIE.
PROPRIETAIRES DE CARRIERES.
FOURNISSEURS DU GOUVERNEMENT
A LA FERTÉ S.JOUARRE (S. ET M.) FRANCE
FRENCH BURR — STONE CROSS
TWB
HURSTMONCEUX
F. CHAMBERS
CANTERBURY
BARRON & SON
LTD.
MAKERS
GLOUCESTER
COMPOSITION STONE — BARNHAM
S:O
1868
CROSS-IN-HAND
G. ARTINGSTALL & Co
WARRINGTON
STONE PLATES
THE MEN
DRESSING TOOL
MAKERS
M.B. 10/77

water-mill in the old parish of Clayton. This was run by the same family that worked the tower-mill mentioned above and the post-mill close by.

During the summer of 1975, Hammond's water-mill was completely demolished, but the pentrough has been saved and is to be incorporated in the restored Ifield water-mill near Crawley. This has only been possible as a result of a dedicated effort of Mr. E. W. Henbery, who has won much support both financial and enthuiast.

After William Cooper's death in 1876, his wife Harriet carried on as head of the firm. She is shown in *Kelly's Directories* from 1878 until 1882, during which time Holloways began to take on most of the business in this part of Sussex and are the subject of the next note but one in alphabetical order.

E. HOLE & SON OF BURGESS HILL

This old family firm is widely known today for work in wind and watermills. Although Holes have had a long association with the scene, they were originally engineers and founders. The business as mill repairers has grown here, whilst the traditional Sussex millwrights finally waned and retired.

Mr. Ernest Hole, the founder of the firm, and father of the present Mr. Edwin Hole[1] (Plate 180), worked in his younger days at Patcham forge, and in 1885 was involved in making the ironwork for Patcham tower-mill, under construction at Mill Hill. This was the last to be built in Sussex.

Some years later in 1892, Mr. Hole set up in business at the forge premises of a small brick-clay working in Turkey Lane, Burgess Hill (Plate 179). The main work was with steam engines, dealing with overhaul and repair work of traction and ploughing engines, stationary steam plant and farm machinery. Business was profitable, and by 1900 the firm was established as 'Iron and Brass Founders and Engineers'. About this time, Mr. Hole was involved in the dismantling of Hooker's steam flour-mill at Twineham (see page 173). The present Mr. E. Hole, who was born in 1905, recalls his father speaking of this and the fact that it had a small windmill on the roof and an undershot water-wheel. A peak runner stone from this mill may be seen at the yard. During the First World War the firm made munitions as well as carrying on steam work, as 'Mid-Sussex Iron & Brass Foundry, Steam Engineers & General Smiths'.

Both Ernest and Edwin, who started work in 1922, were of an inventive nature and involved in many new projects such as potato sorters, earth sifters, self-tipping farm wagons, and a patent mole drain plough. These were no doubt scrutinised by other members of this large Sussex family, principally Mr. Hole of Albourne who was a prosperous local farmer. Another member, Mr. Sydney Hole, who owned 'Hole & Davigdor Dairies', worked with Ernest Hole in the

1930s to perfect an original electric milk float. Other successes were a brick-making machine (Burgess Hill was well known for its bricks and pottery) and rack saw benches. During the last war foundry work ceased altogether, and the firm made many of these benches for the Home Grown Timber Production Department.

From the early 1930s the windmill work began to gain, when, working an Oldland post-mill at Keymer, they repaired the structural timbers and later renewed two sweeps and a stock. Repair and renovation work came their way over the years, and by 1950 was a serious part of the firm's business, both for wind- and water-mills all over England. Outwood post-mill just over the Sussex/Surrey border was one of the early major jobs for Hole and Son, when in August 1952 they replaced the oak weather beam and carried out repair work there.

The following is a summary of the major work concerning Sussex windmills up till the time of writing:

1953 Clayton Post mill—strengthening and shoring-up of the body, plus two new sweeps.

1957 Halnaker tower-mill—renovation and new sweeps.

1958 Shipley smock-mill—complete overhaul and renovation of body, cap, stocks and sweeps. (Norman and Son of Burgess Hill assisting with work on the body.

1960/1 High Salvington post-mill—complete renovation and structural bracing, plus new stocks and sweeps.

1966 Polegate tower-mill—repair work to the cap and fanstage, plus a new set of stocks and partially-shuttered sweeps.

1971 Rottingdean smock-mill—internal steel bracing and new sweeps.

Other mills worked on during this time include those at Chailey, Clymping, Earnley, West Blatchington, West Chiltington, and Winchelsea.

At Argos Hill post-mill, Mayfield, Holes have continued the work of Neves (page 157) and Sands. In 1969 they restored the fan-tackle and fitted new sweeps made by Hole's resident sweep-maker, Mr. Frank Agate (Plate 181). Unfortunately, lightning struck one of these and split the stock. The Wealden District Council have now taken the added precaution of having lightning conductors fitted. During a gale in October 1976, another of the stocks was broken near the canister, highlighting the need for a reliable source of good long leaf pitch pine. After considering the very poor record of substitutes, it was decided (as at Cross in Hand) to fit fabricated steel stocks where possible in future. Early in 1978, Holes made a new one to their own design bearing in mind that its weight should not exceed that of the original. This has now been put up and the mill is back to its previous cheerful appearance. Likewise the construction of a fabricated steel stock in July 1978 has given new life to Meopham smock-mill.

Nearer at hand on Clayton Hill, the tower-mill, Jack has had a fine set of stocks and sweeps hung. (The first since 1910.) These were worked on by Mr. Peter Stenning, and display a respect for traditional millwrighting with a workmanlike form and weather.

Today the yard at Turkey Lane is steadily undergoing modernisation whilst retaining a nostalgia for those who seek it. Plates 182 and 183 show the original 'brick works' forge which dates from the 1870s, and a fine timber jack made in Lewes for Edwin's father.

Work continues both as local engineers and country-wide mill engineers. It is still quite usual to find a sweep being assembled alongside a rack saw bench and a dozen erring lawn mowers. Adjacent to these, small press parts for a local printer are often being turned out, and in the forge, a piece of wrought iron-work, to enhance someone's home or garden.

Today, work on wind- and water-mills is a full-time part of Holes's business. This is no doubt comforting to Tony Hole, who is the third generation here, and with 11 years work with his father, has a full order book for the future.

HOLLOWAY OF SHOREHAM

This firm of Sussex millwrights had a longer and more active career than I at first suspected. Mr. James Holloway was first shown in *Kelly's Directory* in 1866 as Millwright of Ship Street, Shoreham. Business must have been very favourable, for he had moved to Brunswick by 1870 and again to New Road by 1874. There they settled for some 15 years, during which time they were most active with windmills, although the firm built up a considerable reputation as general engineers. By 1895, to cope with the further increase in trade, James Walter Holloway acquired additional premises in Charlton Road, Steyning, and had moved once more to Lower Road, Shoreham, which was to remain the head office. The entry in *Kelly's Directory* for 1905 shows just how celebrated Holloways had become:

> J. W. HOLLOWAY & SONS MILLWRIGHTS & ENGINEERS. Boiler makers and smiths—agents for all kinds of agricultural and other machinery at makers prices—motor car and motor cycle makers and agents. Lower Road Shoreham & Charlton Rd Steyning.

Thus they continued until 1922[1] with the further acquisition of premises at Market Place in Steyning. However, they gave up millwrighting soon after and worked as engineers from 1930[1] or so, until the mid-1950s, whilst carrying out a small amount of mill maintenance.[2] What was left of the trade in Sussex was to be shared by firms such as Neves of Heathfield, E. Hole of Burgess Hill, and others from outside the county.

An early reference to their work with mills dated 1877,[3] mentions their involvement in a mill removal at Worthing. Here, Mr. Holloway supplied a traction engine to assist Mr. Isted in the transportation of Cross Street post-mill on a trolley across Worthing to the eastern seafront. Apparently, a large team of horses had been employed for the initial stages of the move, but, as had been found before, the pull was not steady enough. Sussex oxen would have been employed at one time. However, the steam traction engine was, by that time, in general use. Like many engineers of the time, Holloways hired out traction engines and steam-ploughing equipment. Thus the final stage of the move was accomplished and the mill was subsequently known as Isted's mill.

Cited below are the three best-known examples of J. W. Holloways' work. The first and most renowned in retrospect, was the fitting out of Waterhall tower-mill at Patcham (see page 83) in 1884/5 (for £1,100).[4] This subsequently turned out to be the last corn or provender windmill to be built in Sussex. A great deal of modern ironwork was employed in this mill, and it followed a basic Sussex pattern with a generally similar line to older tower-mills in the area, with a white cap and cement-faced tower. The sweeps were controlled by a variation of Cubitts patent system. Cheale of Lewes perfected the 'spring patent' which allowed each sweep a little independance of the other three whilst retaining Cubitt's 'automatic' overall influence. Holloway wished this mill to be as efficient as possible and had this sytem fitted in a modified form. His 'spring spider' can still be seen today. Holloways do not seem to have used this method when converting Barnham tower-mill in 1890 (see page 74). They did, however, use their unusual screw-down brake mechanism, also employed at West Blatchington and Patcham. Barnham mill had been working for 63 years by this time, and the millwrights completely re-fitted her. The tower was first heightened, enabling longer sweeps to be used. These were standard patents operated conventionally and replaced a pair of commons and a pair of springs. As at Patcham, this tower-mill was well equipped and fitted with three pairs of stones. The iron windshaft is very unusual in that the canister-end was cast separately with a large flange. This was machined and bolted to a complementary fitting on the rear two-thirds of the shaft. The only reason that I can give for this arrangement is that Holloways perhaps didn't have the machinery capacity at the time for finish-turning a single piece of the length required.

At Selsey tower-mill, Holloways were once again employed to refit an old mill. Here, however, their role was a little different in that the tower was completely gutted and re-fitted for intensive grist working, without stones. This was in about 1908, and must have been the last major piece of windmill work carried out by the firm. Inside the cap the cast inscription 'HOLLOWAY ENGINEER SHOREHAM' is to be found on the brake wheel. This cast-iron wheel is similar in detail to those of Barnham and Patcham, but the method of mounting on the windshaft differs in each case. At Selsey the wheel is split and bolted and located on a square shoulder with wedges. Barnham's brake wheel is also split and bolted, but accurately located on a turned shoulder and held with a large steel key. The oldest of the three at Patcham is the most elaborate. Here, the split wheel's large square aperture was packed and wedged on to a cast centering box which was in turn secured to a square shoulder on the windshaft. The reason for these variations is likely that Holloways used an old windshaft at Patcham as they were to later at Selsey. ·This was a trade where many pieces of machinery were retained or passed from mill to mill as is well illustrated in the note concerning Neves, a little further on.

MEDHURST OF LEWES

Samuel Medhurst was first registered as a millwright in 1824[1] and shortly afterwards at the age of 26 set up with his brother William, then 28, in the High Street of St. Annes, Lewes (now Western Road). His work on Sussex windmills and water-mills was prolific and the name Medhurst became synonymous with the best and most up-to-date methods in mill practice. The two brothers who were of milling stock, had been involved in the removal of Town mill from Pipe Passage to a site just north of the prison wall, where came to be known as Smart's and then Shelly's mill (Plate 135). This is credited to William Medhurst and the researches of Mr. H. E. S. Simmons have put this at *c.* 1818. This, the earliest reference to Medhurst's work, is the only one that I know of concerning actual smock-mill construction, save that Samuel was employed in 1853 to raise the same mill 20ft. or so at a cost of £150.[3] Apart from this, their windmill work in Sussex appears to have been almost entirely on post-mills. It was a considerable surprise to me that there is no direct evidence of an original complete Medhurst windmill and that they would certainly have been too young to be involved in work between 1805 and 1814[4] often credited to them. I suspect that they were responsible for the impressive Ashcombe post-mill on Kingston Down (Plate 104). This mill was a sophisticated hybrid erected in 1832 for Mr. John Weston, also of St. Annes, Lewes. A unique triple-canister windshaft which was cast at the Pheonix Iron Works in Lewes, carried six double-shuttered patent sweeps. These appear rather flat at the tip and were, originally, equipped with seven bays, five with three large shutters in each, and the two outermost with five and four narrow shutters respectively. Medhursts were undoubtedly responsible for these sweeps as they held the licence for Cubitt's patent in these parts. The fitting of these to many mills, post-, smock- and tower-, was the greater part of the business at one time. In any case, the erection of such a unique mill on their doorstep by a competitor would have been an unbearable affront to this already famous firm.

John Every, who founded the Phoenix Iron Works (latterly East Sussex Engineering) on the banks of the Ouse in 1832,[2] worked very closely with the Medhursts and their collaboration produced well-designed and utilised iron-work which is still evident today, particularly so at Cross in Hand post-mill which was rebuilt by Samuel Medhurst in 1855 and moved again in 1868 by him. The crown tree bracket bears the words S MEDHURST MILLWRIGHT LEWES JUNE 1855. At the time of the second move they fitted the Sussex tailpole fan-tackle often attributed to them. This arrangement, as can be seen today, placed the 11ft. ∅ fan well for wind at the outer end of the tailpole (see page 23). Medhurst fitted this to a number of other post-mills, including Argos Hill, where the detail of some of the ironwork is identical to Cross in Hand. Golden Cross, which was re-fitted by Medhurst after storm damage in the late 1860s, was equipped with tailpole fan-tackle at the time. I think it very probable that the firm added this feature to a number of post-mills when converting and modernising with patent gear. Jill at Clayton and Sutton mill at Seaford are other examples. When converting a mill to Cubitt's patent, a new iron windshaft was usually incorporated with a cored hole through the centre for the striking rod. Several of these were rather slender for their length and prone to undue vibration when under load. So much so that the unusual precaution of a third steady bearing was resorted to at Cross in Hand. Likewise at Hurstmonceux, which was re-fitted by the Medhursts. This problem was overcome with a stout longitudinal bracing of clamped oak planks between the brake and tail wheels.

This Lewes firm of millwrights prospered under Samuel's guidance (William does not appear to have been involved after 1845), until around 1870, when one of his sons, Boaz, took over. The illustration (Plate 184), from the Reeves collection, and dated 1869, shows 70-year-old Samuel with his granddaughter centre and Boaz extreme right. The new water-wheel for Fletching mill had just been completed. Unfortunately, Boaz died in 1878. The heavy iron windshaft at Keymer post-mill has his mark B MEDHURST LEWES 1873. Similarly, at Hurstmonceux post-mill, we find B MEDHURST MILLWRIGHT & ENGINEER LEWES SUSSEX on a bridge iron.

Within the next ten years, all of the Medhurst menfolk left in this family business died. Samuel's brother, William, followed Boaz in 1878, then younger son, Mark, at the age of 38 in 1881, and lastly, Sam himself in 1887 during his 89th year. Today in St. Anne's churchyard a rather sad line of cast-iron grave markers (Plate 185) provided by John Every, is headed by that for Samuel and his wife, Philadelphia. They had eleven children, and Samuel outlived all but one, a daughter. A little later, John Every joined his life-long friend.

The business was then taken on by Alfred Shaw until 1903, followed by Bishop and Wells until 1918. Some 10 years later, after 100 years or so of use as a millwright's yard, the site at St. Annes was bought by Lewes Motors, who built the large garage that can be seen today.

The post-mill at Cross in Hand is to be restored. A fitting memorial to Samuel Medhurst, it sums up the best of his windmill work: rebuilt as a very large post-mill with two pairs of spur gear-driven head-stones, and a third pair of overdrift tail-stones, all driven by patent sweeps, and the whole winded by Sussex tailpole fan-tackle.

F. NEVE & SONS, HEATHFIELD

This firm, well-known as millwrights and engineers for over 100 years, started in Flitterbrook Lane, Warbleton,[1] as Stephen Neve. Mr. S. Neve had previously worked for Upfield. *Kelly's Directory* for 1855 shows Thomas Upfield as millwright in Warbleton. Stephen Neve appears to have taken on his yard, for it is he who is shown here from 1858[2] till 1882.[3] His sons, Henry and Frederick, took on the business and were registered here in 1887.[3] A little later the brothers moved to Station Road, Heathfield, and worked from 1895 as Neve Brothers—Machinists. Approximately 15 years later these two split up, and Frederick took the millwrights' business to the High Street, leaving Henry to continue as 'Neve Brothers—general engineers' in Station Road. Frederick continued as 'Frederick Neve' and then F. Neve and Sons.

The First World War left very little profitable millwrighting and it is surprising that they were able to continue, but they did, as the ledger for 1921-30 quoted a little further on shows. Latterly, the sons, Jack and Walter Neve, worked in close association with William Sands and Son of Heathfield, and were the last of the county's traditional millwrights. The yard closed after Walter's death in 1961.

Neves were responsible for some of the smock-mills built in East Sussex (mostly in association with Upfields, who worked from Catsfield in later years). These were very well planned and executed, reflecting the Kentish style, often having characteristic Kentish-hooded caps and fanstages. The stages differed from the Sussex pattern, being set somewhat higher and well placed for wind. The supporting timbers for the fan were attractively set at an angle and well cross-braced.

Neves were also concerned in the building of one Sussex tower-mill —Stone Cross (page 89). The mill here was erected in 1876. Stone Cross mill has a perfect Sussex beehive cap, Kentish fanstage, and Kentish rocking-lever system for striking the patent sweeps. The tower was contracted to a builder who had the round windows fitted. The whole mill has a fineness of feature which the tower-mills of West Sussex cannot lay claim to.

This firm was concerned in the maintenance and restoration of windmills for many years and after the final decline were notably involved in the work at Chailey smock-mill (page 58) and Halnaker tower-mill (page 78) in the mid-'thirties as F. and W. Neve, and West Blatchington just before the war.

To illustrate the long association of a firm like Neves with the Sussex windmill scene, I include the following note of a journey taking 77 years and made from the Weald of Kent at the start of Mr. Stephen Neves's career as journeyman millwright, to the centre of Sussex. The thread is perhaps tenuous, but the reader will, I am sure, forgive a little licence here.

We take up the story at Biddenden in Kent about four miles from Cranbrook, the home of the finest smock-mill in England, and near Sissinghurst, which takes its name from an historic fortified manor close to the village. By autumn 1859 the smock-mill here was becoming redundant and likely a liability to the owner. The windshaft was still, the sweeps left struck up. Cherry Clack mill as she was called, perhaps had a hard time competing with the towering lady of Cranbrook and other mills close by.

Some 20 miles away in East Sussex, high on the Wealden Ridge above Punnett's Town, Blackheath post-mill had recently burned down, as re-told in the following account:

Sussex Advertiser, 7 September 1859.

Blackdown Windmill destroyed by Fire. On Thursday morning last, the 1st. inst., about 3 o'clock, this mill belonging to Mr. S. Dallaway, and in his occupation, was discovered to be on fire, and in a little more than an hour the whole was entirely consumed, including a new round-house under. The mill and round-house contained about 3 loads of wheat (ground and unground). Ere alarm could be given and assistance got, the 'devouring element' had destroyed nearly all, besides a considerable quantity of other corn etc., a small portion only being rescued. The fire, it would seem, originated as follows:—Mr Dallaway and his son (or sons) were at work in the mill till late on Wednesday evening (Mr D leaving first). At about half past 10, there not being much wind, his son (the grinder) decided on going indoors to bed. He stopped the mill and snuffed out the candle (it is supposed with his fingers) and the son thinks the snuff must have fallen into the oat bin, into which the meal had just been running, where it lay on the top smouldering and coaling until it engendered flames, and was discovered by some one about 3 o'clock. Mr Dallaway's cottage, where he resides, is close by the mill, and it was not without great energy that this was prevented from catching fire. Mr D had insured up to about the time of taking the premises in his own occupation (about 3 or 4 years ago), but somehow or other he has unfortunately not since insured, and consequently the whole of this, to him most severe, not to say ruinous loss, falls on him, unless public sympathy can be enlisted on his behalf.

It was not long before Mr. Dallaway and the local millwright, Mr. Stephen Neve, came up with the solution. 'The smock mill near Sissinghurst is standing idle, we could bring her over'. 'Sounds alright, we can rebuilt her up here and perhaps make use of some of the old timbers left from Blackheath mill as well'. An offer was made and accepted.

The millwrights dismantled Cherry Clack mill late in 1859, and Mr. Hobden, a relation of the Dallaways, transported the whole structure on large broad-wheeled wagons along the Wealden Ridge to Punnett's Town, drawn by horses perhaps when the going was easy and flat, then by oxen when a stronger, steadier, albeit slower pull was needed. (The self-propelled traction engine had not truly arrived yet.) This must have taken several weeks, and the cost of employing men away from home mounted steadily. Soon after the arrival at Punnett's Town the structure was started. A base was raised, built of sandstone quarried nearby, and the rebuilding of the smock-tower began. Here, four of the sides were usable and matched with new cant posts and framing to make the whole. The machinery was installed, using derrick poles and scaffolding. Next, the cap framing was rebuilt on the newly-laid iron curb. It was decided that the old windshaft was not suitable, perhaps because it was damaged, or that the wooden shaft was not drilled for patents. The iron canister-end of Blackheath mill had survived the fire and so it was that this, with a new iron tail strapped and wedged to it forming a complete shaft, was placed in the mill and the brake wheel fitted. A horse- or ox-winch would have been used to raise a heavy piece such as this.

The charred tailpole from the old post-mill had been placed in the new base as a floor beam, and the date MDCCCLIX[5] incised in its side. Rebuilding the whole mill would have taken several months, and I should think that grinding started again in early 1860. 'We'll call her "Punnett's Town" Cherry Clack Mill', they said, 'That'll be right and proper'.

Here, on this lofty spot, she worked for 70 years, maintained and cared for by the successive generations of Neves and Dallaways until the late 1929. The windshaft became still once more and the power-mill close by took on the work.

On a January day in 1928 a great storm had raged causing a great deal of damage throughout Sussex, and many windmills were battered by its force. At the centre of the county, Chailey smock-mill suffered badly. The whole of her cap and windshaft assembly was brought to the ground and almost completely demolished. F. Neve and Sons were consulted, and the cost of repair found to be prohibitive. Here the smashed remains lay until the building of a new block for Chailey Heritage made repairs possible by public subscription in 1933. Sadly, however, the old iron windshaft had been weakened by its previous fall, and in a storm in the autumn of 1935, sheared in front of the brake wheel bringing the canister-end and the sweeps crashing to the ground once more.

Fifteen miles east at Punnett's Town, Cherry Clack mill stood decapitated and forlorn. She had died (temporarily as it turned out) in late 1935 when Neves were employed to dismantle the mill for scrap. The windshaft was removed along with the other machinery, and Neves were sensible in keeping it for our story would have no end but for their foresight.

In 1936, when Chailey mill needed a new windshaft, patterns for a new casting did not exist. So it was that the final journey for the windshaft began. The cannibalised shaft (Plate 186) was brought over from Heathfield and fitted through a specially cast dummy brake wheel, and the whole securely bolted to the cap sheers. This rests here today, a tribute to the well-known Sussex millwrights it has outlived.

Mr. Sidney Ashdown has been kind enough to allow me to make the following extracts from one of Neves's ledgers in his possession.

During the early 1920s, windmilling picked up a little after the crippling effect of the First World War and many repairs to mills are recorded in Neves's ledger for 1921–30.[4] A great deal of work on water-mills and power-mills is also found. However, although new work for several power-mills is recorded, this could not prevent the final decline. The windmill at Punnett's Town was stopped due to bad troubles with the curb; the sweeps were finally removed in 1929. The little post-mill at Blackboys was later fitted up with these after repair work and modifications. Chailey mill appears to have had a lot of work on her fan and curb before the final storm damage in 1928. Captain Hardy of Mayfield had a number of repairs carried out on his mill at Argos Hill during this time, starting with the Sussex tailpole fan-tackle:

Hardy Captain *Argos Hill Mill*

1921		£	s.	d.
Oct 13	To making inspection of tailtree of mill taking necessary measures for repairs		10	8
22	To preparing timber and bolts at shop + fitting and fixing new timbers. Refitting fantail at mill, labour and power	7	18	9
	Men's travelling expenses		10	6
	New bolts for braces, large nails, paint, washers		10	4
	3 large bands & hasps complete	2	9	6
	2 large planks .. .. @ 18/6	1	17	0
	Use of jacks, pulleys & tackling for the work		3	9
	To carriage of materials to mill		6	0
		14	6	6

1922				
Jul 11	To inspecting fantail gear & preparing pattern for new worm to travelling gear	1	2	0
Mar 18	To fitting worm to fanshaft & making new bearings for fantail shaft— @ shop labour & power	1	9	6
	Self and man to mill overhauling fantackle, adjusting & tightening gear & fixing new worm	2	6	9
	Rail fares		3	9
	New worm casting + 2 bearings, set screws		12	9
		5	14	9

		£	s.	d.
1924				
Jan 12	To welding on new end to fantackle shaft, turning same + fitting & pinning on worm with steel pin *Breakdown job* + carriage to Mayfield	1	10	4
May 31	To overhauling the fangear & bearings of Argos Hill Mill Making & fitting new iron work & refitting Bearings shafts etc.			
	Time at shop & mill	2	15	11
	17½lb forgings & bolts		5	10
	Washers, Bush for shaft, long nails		2	3
	Use of chisels, files @ 1/6 travelling 4/6		6	0

This appears to have cured the trouble and we find work on the internal machinery.

1924		£	s.	d.
Jun 7	To raising & parking the windshaft, adjusting gears & brake of Argos Hill Mill. Time + materials required	2	9	0
	Use of large screws & tackling		7	9
	Men's travelling expenses		3	0
		8	0	1

The detail of driving gear repairs here was charged to Mr. W. J. Richardson the miller. These were for what was called tenants fixtures and fittings.

Richardson W J Argos Hill Mill

1924		£	s.	d.
Apr 12	To re-forging horns of quant, trim step brass, foot of spindle, time & power for lathe, coals, files etc.	1	10	4
May 31	Shanking 12 cogs for brakewheel & repairing gear at mill—time, materials, power & travelling expenses	3	0	0
1925				
Jan 10	Repairing gear in tail wheel		5	3
Mar 14	1 pair of governors complete	1	5	0
Apl 18	.. ..		8	0
Sep 19	Preparing 12 cogs for brakewheel & repairing gear at mill. Fitting iron plates on stone joists. Adjust windshaft. Time & power 1/6 Travelling 2/6	2	9	0
	Stuff for cogs 12 @ 4d		4	0
Nov 14	Man repairing gear of brakewheel 5½ hrs & rail		10	8
		9	12	3

Again Captain Hardy is charged for structural work:

		£	s.	d.
1925				
March	To fitting strong angle irons to corner breast posts of mill. Packing & bolting side beams & bushing hole in roundbeam (windshaft) as per estimate	13	10	0

This mill finally stopped work in 1927.

The reader will have noticed how little these repairs appear to cost, but we must reflect on the changing fortunes of the pound in the last 50 years.

Below are examples of the cost of stones. These were very expensive items. Mr. Hemsley's £62 representing perhaps £700 today.

Ellis Bros *Summerhill Mill Warbleton*

		£	s.	d.
1923				
Mar 3	1 Best quality Derby peak runner 4ft 4ins diameter, faced hooped, eye cut out and furrowed	16	19	0

Stevens R *Earnley Mill Chichester*

		£	s.	d.
1924				
Jul 30	1—3ft 10ins diameter composition Burr Runner Dressed 'with Sun'— Delivered Chichester station	13	8	9

Dallaway Bros *Punnetts Town Mill*

		£	s.	d.
1925				
Sep 19	1 pair 52in aloxite composition stones ready for work furrowed & faced	36	15	0

Hemsley & Son *Isfield (Watermill)*

		£	s.	d.
1925				
	2 pairs 48ins diameter emery & french burr composition stones furrowed 'against the sun' & faced for work @ £31 per pair	62	0	0

McDougalls Ltd *Horsebridge (Watermill)*

		£	s.	d.
1927				
Dec 17	1—44ins diameter composition, carburundum & emery runner stone furrowed & faced for 1ork £18 less 2½%	17	11	0

Another series of entries shows work carried out for David Catt and Son of Hamlin smock-mill, Hailsham. The Catts were very well known as millers in East Sussex. One William Catt ran the equally famous tide-mills on the flats behind Bishopstone beach for many years. This huge mill worked 16 pairs of stones and was topped by a small smock-mill, used for hoisting purposes only. This almost feudal complex stopped work in 1883 and was razed to the ground during the last war. The Catts of Hailsham fared better, working Hamlin's mill. Neves ledger shows work done on the sweeps and machinery inside here. Unfortunately the mill was burned to the ground late in November 1923 and we find an entry for that month: 'Surveying for new mill'. During the following year over £600 worth of work is entered on a new power mill for Mr. Catt. This worked until 1967 when Mr. Catt finally closed down, not long after Neves themselves finished work as millwrights.

Extract from a Draft Agreement for Building a Smock Windmill

Finally, in this chapter I have included the following memorandum for building Frenches mill which stood at Croft Road, Hastings. I am indebted to Mr. Purdey of Hastings Reference Library for help in securing this and other valuable information from the J. M. Baines collection:

Dated April 20th 1799.

Between Cornelius Wetherell of Robertsbridge, Millwright & John French of the Town & Port of Hastings, Baker.

Cornelius Wetherell in the consideration of £300 . . . to erect, build and completely finish . . . fit for working by 25th September next upon a piece of land near the Fighting Cocks in the Parish of St Clements, lately purchased by John French:

A Smock Windmill

To contain at the bottom 18ft from outside to outside and 10ft 3ins at the top.
In height from ground floor 40ft. To consist of 3 floors: Bottom, Middle and Top.
The brick work (Base) to be carried up 1ft 6ins from the ground.

The Smock

–The botton floor to be of fir girders 1ft by 1. The joists of the fir 4ins x 3. To contain in height 9ft.
–Eight corner posts 30ft long 9ins by 9. Eight oak plates for the bottom curve 9 ins by 4½.
–The middle floor of 2 fir girders 1ft square and the joists of fir 4ins by 3. To contain in height 9ft
–The top floor of 2 fir girders 1ft 4ins by 8ins To contain in height 10ft. The top curve of elm, 10ft 3ins diameter, 9½ins by 8.
–Floorboards, weatherboarding & nails

The Cap

–Pieces for frame 6ins by 5
Punchins and braces 6ins by 4.
Quartering 4ins by 3.
Sliding curve of oak 8ins by 6.
Rafters of elm or oak 3ins by 2½.

The Machinery

–The roundbeam of oak 2ft at the largest end and 1ft 3 ins at the smallest.
–Brakewell 8ft. 3ins diameter.
–Wallower nut 3ft 6ins diameter.
–Upright fir shaft.
–Spur wheel 5ft 2ins diameter.
–Spur nut for the French stones 1ft 8ins
–Spur nut for the Peak stones 1ft 6ins.
–Spur nuts for Sack Tackle and machine
–Cross gudgeons and brasses
–One pair of peak stones 4ft 4ins.
–One pair of French stones 4ft 4ins or 4ft 2ins.
–Cast iron quants, spindles & horns.
–Bridge trees & carriage.

—Tuns & bins
Meal troughs & spouts
A 6ft machine (dresser).
Sack tackle & machine ropes.
Riggers & stone boxes.

The Sweeps
Stocks of fir 36ft long 11ins by 8.
Sail Rods of fir 32ft long 8ins by 5.
Sail Cloths 4ft 6ins wide.
Scantlings.

Winding Gear
—Worm & Wheel with Rope.

The said John French to pay the sum of £300 in the manner following: The sum of £50 paid when the timber shall be but on the ground—the sum of £50 within one month after and the remaining sum of £200 when the mill shall be complete. Each of the parties bindeth himself in the penal sum of £50 firmly by their Presence.

As Witness—their marks.

This was a relatively inexpensive mill (almost entirely constructed of timber), hand winded and driven by common sweeps.

Mr. French worked the mill from 1799 until his death in 1820. In *c.* 1838 this was taken down after several years of inactivity and re-erected at Silverill, where she was worked until 1862.

Mr. Draper of Silverhill, whose family worked the business in latter years, described the mill as, very primitive almost all wooden, fitted with one pair of peaks and one pair of burrs, two common sweeps and two spring sweeps (converted at sometime). The mill was re-modelled in 1862 using iron machinery and fitted with three pairs of stones worked until 1867, when it was burned down.

Upfields, the well-known local millwrights, built a larger new smock-mill on the site shortly after. This came to be known variously as Silverhill, Drapers or Tivoli mill, standing until July 1966. Towards the end, great concern was expressed at her declining condition, but as with so many others, demolition was the final outcome.

Drapers mill was the last to stand in the Hastings area. Nothing remains at the site today. The base of Frenches mill was subsequently used as foundation of part of a house at Croft Road, Hastings.

CHAPTER NINE

APPENDIX I

OBSCURE REMAINS

FOLLOWING is a list of mill remains known to the author which can only, in the broadest sense, be classed as 'obscure' today. I very much hope that any reader who knows of others, or, indeed, has any information to pass on, will not hesitate to do so. Any useful note or reference will be included at a later date. These are listed alphabetically by location and type.

POST MILLS

ARLINGTON (Windover Post-mill)

Traces of roundhouse walls and a millstone under the turf at Windover Hill. Built early 1800[2]; two pairs of stones, tailpole and talthur, two spring sweeps, two commons, burned down in 1881–2. Known as Windo'or mill.[1]

[TQ 537-035]

BLACKBOYS (Post Mill)

Foundations of the brick piers in the grass of mill field. Built 1807 on Glynde Hill, moved to Blackboys in 1867–8. White mill; two pairs stones, tailpole and talthur, spring sweeps, pulled down in 1945.[1] **[TQ 520-206]**

CROWBOROUGH (Beacon Post-mill)

The pintle and lower portion of the post are exhibited at Argos Hill mill. Built 1782; two pairs of stones, tailpole and talthur, white mill, latterly tarred and embellished with a look-out stage, burned down 1940/44, cleared 1968.[1]

Site of remains [TQ 571-283]
Original site [TQ 513-309]

FISHBOURNE (Post Mill)

Foundations of the brick piers in the garden of mill house. S. and E. just visible. Moved from Rustington, October 1857,[2] to Fishbourne. A black open trestle mill; two pairs of stones, tailpole and talthur, four commons. Pulled down summer 1898.[3] [SU 838-044]

GLYNDE (Hollow Post Pumping-mill)

Remains of timbers and machinery engulfed in a bush close to railway line (Plate 187). Mid-19th-century hollow post-mill; single-throw crankshaft, gear driven from above, two lift pumps, long-boarded sweeps. Used, I understand, to supply water to steam engines which drove an aerial ropeway for a nearby quarry.

[TQ 457-087]

HEATHFIELD (Mutton Hall Post-mill)

Iron fintail canister end of windshaft (Plate 188). (To be taken to Park mill, Burwash.) Built 1805; two pairs of stones, tailpole and talthur, spring sweeps. Pulled down in 1891.[1] Original site [TQ 581-218]

IFIELD (Post-mill)

Remains of trestle timbers are to be seen in Goff's Park, Crawley (Plates 2 and 3). Mill said to have been moved from Bletchingly *c.* 1870. Very

unusual box-shaped body, roof pitched almost from the side girts and tapered down to tail. Three-storey roundhouse, tailpole and talthur, patent sweeps, demolished 1899.[1] Site of remains [TQ 260–363]
Original site [TQ 251–379]

LEWES (Southern Post-mill)

Brickwork and foundations of roundhouse incorporated in a swimming pool. Built *c.* 1785; white mill, tailpole and talthur, four spring sweeps controlled by Cheale's 'spring patent' (Plate 8). Demolished 1912/13.[1] [TQ 402–093]

LEWES (Spital Post-mill)

Foundations of brick piers reduced to ground level (difficult to see now) a granite boundary stone markes the site. Built 1770; white open trestle-mill, named after St. Nicholas hospital nearby. Burned down May 1885.[1]
[TQ 399–103]

SHOREHAM (Good Friday Post-mill)

Barest trace of flint-walling, part of the roundhouse, are to be found in a garden hedge. Built *c.* 1764; black mill; tailpole and talthur. Destroyed by fire 1899.[1] [TQ 212–066]

SINGLETON (Open Air Museum, Hollow Post Pumping-mill)

Reconstruction of a pumping-mill, which stood in Westham, close to the station. The post and machinery are original. Erected here in 1975 (Plate 189). Mid-19th-century hollow post-mill; eccentric cams driven by upright shaft, two lift pumps, cloth sweeps, winded by vane. Used to drain brick clay workings.[4] Hollow post pumping-mill [SU 874–128]
Original site [TQ 640–043]

TWINEHAM (Hooker's Wind- and Water-mill) (Hollow Post mill)

Remains of water-mill building that once supported a hollow post-mill. Built *c.* 1865,[1] believed to have been similar to that at West Ashling. Dismantled 1900.[5] (see page 173) [TQ 258–203]

WARBLETON (Summer Hill Post-mill)

Foundations of the brick piers in a farmyard. Built 1824/5; white mill; two pair of stones, tailpole and talthur, patent sweeps. Pulled down August 1935,[1] roundhouse cleared 1974. [TQ 619-191]

WEST ASHLING (Wind- and Water-mill)

Water-mill building that once supported a hollow post-mill. Built *c.* 1860[1] by Armfield; white mill; fan-tackle patent sweeps drove three pairs of stones. Cleared 1940 and finally taken down in 1955. Mill building becoming derelict today (Plates 197 and 198, and see page 173). [SU 808–073]

SMOCK-MILLS

BEXHILL (Pankhurst's Smock-mill) (Sidley)

Foundations of base walls cleared in 1975. Built *c.* 1798; white mill; two pairs of stones, fantail, spring sweeps. Taken down in 1928 and removed to Stocks Green in Kent, given new sweeps and tarred.[1] Collapsed September 1963.[4] [TQ 732–095]

HASTINGS, Croft Road (French's Smock-mill)

Part of mid-19th-century house[7] built on the foundations of base. Built 1799[8] (see page 163). Two pairs of stones, hand winded. Common sweeps. Dismantled and taken to Silverhill *c.* 1838. Re-worked using iron 1862. Destroyed by fire 1867. [TQ 824-102]

WALBERTON (Short's Smock-mill)

Part of the base remained up till latter times; however, this is not at all evident today.[6] Built *c.* 1820. Badly damaged by storm in 1887; taken down *c.* 1896.[1]

[SU 967-062]

WINDWHEELS

PUNNETT'S TOWN

Remains of a carpenter's shop which supported a windwheel. Built *c.* 1908. Wheel taken down in 1916.[9] Building subsequently converted.

[TQ 623-204]

WORKING MODEL WINDMILLS

Three recent working model windmills are mentioned here. The first of these was built by Mr. Sidney Ashdown of Cross in Hand mills in 1934. This was a small post-mill with four shuttered sweeps and winded by a tailpole fan. Mr. Ashdown tells me that this developed enough power to drive a small generator, if rather sporadically. He was not able to pursue this aspect of his interest for much longer as he became a full-time miller at Cross in Hand mills after his father's death in 1936. A photograph of the model standing in front of the existing post-mill in 1935 is to be found in *Windmills and Millwrighting* by Stanley Freese.

Plate 124a shows the second of two one-third scale model post-mills built by Mr. R. J. de Little, an enthusiastic windmillian and writer on the subject.[1]

During 1957 and 1958, at the age of 14, he erected the first of these, a hand-winded open-trestle post-mill, which started life with four common sweeps, later changed to two commons and two spring sweeps.

The second of his mills was erected in 1959 and more ambitious still. Built to the same scale, this was equipped with six single-shuttered spring sweeps, 12ft. long and controlled by coil springs. The body was winded by a tailpole mounted fan 4ft. in ∅ and the trestle enclosed in weatherboarded roundhouse. Using the traditional method, Mr. de Little morticed the three stocks through the poll end of the wooden windshaft (Plate 191).

Within the body, a wooden clasp arm brake-wheel 4ft. Ø drove a late 19th-century kibbler by belt. The power developed by the mill was more than sufficient for this work and empirically rated at 4 h.p. maximum. As may be seen in the photograph, the fan-tackle was based on the Sussex pattern, with a six-blade fan which was later fitted with more fulsome blades to improve winding.

During its heyday, when time and expenditure permitted, this unique model was periodically used to produce kibbled oats. Sadly, in a freak storm early in 1961, the first mill was blown over and the second suffered damage to the sweeps.

It was decided to re-work the second mill using a new poll-end fitted with two stocks. These carried two spring sweeps and two commons. A Bentall plate mill which ground barley was fitted in place of the kibbler and remains within. The second illustration shows the model today without the fan and boasting only two spring sweep frames (Plate 192).

This stands on private ground a little to the north of Broadhill in Keymer.

[TQ 319-170]

APPENDIX II

BYGONE MISCELLANEA

BISHOPSTONE TIDE-MILLS (Hoisting Windmill)

About half-way round the bay on the marshy flats between Newhaven and Seaford, there once stood the largest water-mill in Sussex. Erected in 1778 and worked by tidal flow, this was, in its latter years,[1] said to have been equipped with 16 pairs of stones variously driven by three undershot water-wheels. Using 20 acres or so of storage ponds,[2] the mill was held to be capable of working at the rate of 1,500 sacks per week, with favourable tides and a 16-hour working day, i.e., two hours before low tide, 12 till next low, plus two hours after and grinding for four to five hours at each low. The mill creek, fed by the Ouse, is situated behind the beach scarp and followed in part the course which the river had once run before storms breached her through at Newhaven, or Meeching as it was then known.

Mr. William Catt headed a somewhat feudal complex here between 1803 and 1853.[3] A small village, including the mills and housing for his workers, sprung up, and the whole faired well. Later in 1864, when the railway came through, fortunes rapidly declined as grain could then be taken direct from the port to large steam-mills inland.

The windmill was probably erected around this time or a little earlier to improve efficiency (Plate 193). Sailing barges passed up the creek, drawn by horses when necessary, to be unloaded. The six-sided smock-mill drove hoisting tackle of various forms, one of which was an endless belt with buckets attached, used to raise grain within. Mr. Stan Tubb, who knew the mills after they closed down, remembers seeing what remained of the thick hide belting many years later. No doubt this little mill was used when the available water power was required for grinding or between tides. She was driven by patent sweeps and winded by a fantail, and mounted on the roof of one of the mill buildings. Whilst not being certain of wind at all times, there would, of course, have been the sea winds which could be relied upon, as could the tidal flow for the wheels.

After Mr. Catt's death in 1853, Bishopstone tide-mills were never to be quite the same, and the coming of the railway put the seal on their fate. In 1883/4 a storm did considerable damage to the water-wheels and the windmill, forcing the complete close down of grinding.

The buildings were subsequently to be used for waterhousing until 1901, when they were partially demolished. During the First World War, a seaplane base was established on the seaward side, followed by a racing stables, and a holiday home for Chailey Heritage.[4]

Finally, in the last war, the army used what remained for various purposes, including weapon training, and then razed the site when Seaford Bay was requisitioned as an assembly point for the 'D'-Day landings.[2]

PUMPING WINDMILLS

To my knowledge, there were at least four pumping-mills remaining in Sussex at the start of the last war: two hollow post pumping-mills, at Glynde and Westham (mentioned in the section on 'Obscure Remains', pp. 166 and 167), a small smock-mill erected at Birdham in the mid-1930s, and the derelict tower-mill at East Blatchington.

BIRDHAM

This was a small black smock-mill with short white sweeps, used for raising water via a crank and lift pump (Plate 194). The firm, Tinsdale of Chichester, erected her for a local farmer in 1935.[1] Whether it was that the source of water was found to be inadequate or that the mill was troublesome, it did not, in any case, survive for long. By 1939,[1] pumping by wind ceased, and the smock was allowed to deteriorate. She was gone by 1946. The site was about half-a-mile north-west of Birdham church.

EAST BLATCHINGTON

Built in 1882 by Warner to raise fresh water for the Newhaven and Seaford Water Company,[1] this proved to be an extravagant failure.

A report in the *Sussex Advertiser,* dated 20 November 1886, gives details of a board meeting of the company when it was stated that the mill was unsatisfactory. This may have been due in part to its rather sheltered position below what later became East Blatchington golf course. Samuel Medhurst, then in his 88th year, was asked to advise. The result was that £270-worth of second-hand machinery was purchased and the mill re-worked. In its final form this mill reflected the Lincolnshire practise of mounting the patent sweeps on an iron windshaft which had a cross as opposed to the usual canister for securing the stocks. The six-storey tarred brick tower carried a white beehive cap.

The wind-power derived by the re-worked mill was to be used to supplement a steam pump near at hand. Apparently, the water table was contaminated with sea water (the well was only a mile from the shore line) and quite unsuitable. A pumping station was subsequently built at Bishopstone and proved more than adequate for local needs.

The tower-mill became a curiosity without sweeps and fan (Plate 195). During 1914[1] the machinery was removed, leaving the empty tower to bear witness to the whole sad episode. This was demolished shortly after the last war.

SAW WINDMILL

PUNNETT'S TOWN

Brought up from a site in Horam, four miles or so the south-west in 1866, where it had been used to drive farm and feed machinery only,[1] this little white smock-mill was used to drive saw benches for the Lower family. She was known as Punnett's Town saw smock-mill, to distinguish her from Cherry Clack mill, which worked at the top of the same lane. As may be seen (Plate 196), the smock was re-erected on a square timber base which formed part of the sawmill. The Kentish hood carried patent sweeps, with curious small additional wind boards at the tips, and a six-blade fantail. In Holland, several wind-driven saw-mills may still be seen, these usually employing vertical reciprocating saw frames. At Punnett's Town power from the upright shaft was taken to line shafting which drove the saws and benches by belt or eccentric crank. Sadly, as with all windmills, however well placed, natural power could not be relied upon. Engine-power eventually took on the work and the windmill became redundant. This unique mill was taken down in late 1933 and the site sold for housing.

WIND- AND WATER-MILLS

TWINEHAM (Hooker's Mill)

A photograph *c.* 1895, in the Gurney Wilson collection shows this to have been a fair-sized water-mill with a windmill base at roof level. The windmill had gone by this time. A notice in the *Miller,* dated 7 February 1876 reads as follows:

> To let Hookers flour mill containing 5 pairs of stones, wind, steam and water. Situated 4 miles from Burgess Hill. Good mill house, smiths shop, bakehouse etc. Apply Mr Wood, Hickstead.

The complex, erected *c.* 1851, was obviously in full trade and working well at this time. Mr. George Packham, a well-known local mill owner, had been apprenticed to a Lewes millwright and had added the windmill for one of his daughters approximately 10 years earlier.[1] She was married to Mr. Wood and leased the mill to Mr. C. Packham of Cobbs mill in 1886. Working by water ceased when Mr. Packham took over. The windmill was damaged a year later, after which steam power was employed for five years or so.

Of the windmill, little is known, save that it was similar to that at West Ashling (described next). The researches of Mr. H. E. S. Simmons indicate that four patent sweeps drove the upright shaft of the water-mill through the hollow iron post of the trestle. A skeletal framework which carried the sweeps was winded by a large fan.

Mr. Ernest Hole of Burgess Hill was employed to dismantle the machinery in 1900. Most of the buildings were used for cattle shelter or storage for many years until they either fell down or were demolished. The dried-up mill stream and part of an outbuilding are still evident at the site (not one of the best for wind), 500yds. north-east of Twineham church. [TQ 258-203]

WEST ASHLING (Funtington) (Hackett's mill)

As with Hooker's mill, this was a supplementary power supply and a hollow post-mill. Plate 197 shows this mill, *c.* 1900, clashing hideously with the original water-mill and its rural surroundings.

Armfields, the Hampshire millwrights, specialised in improving the traditional watermills and erected the windmill on piers built up inside the mill 1859/61.

The stones bore the date 1859, but a local tradition held that the structure was put up during 1861 when the spire of Chichester cathedral fell. The millwrights later fitted a turbine in place of the water-wheel. I believe that this was an Armfields' river type.

The windmill was an unusual hollow post structure, comprising a small half-body which carried an iron windshaft with four patent sweeps. The power was taken from the small headwheel via a wallower and upright shaft down through a hollow wrought iron tube into the main building. Three pairs of stones were laid out below, and, judging by the size of the sweeps, it may well have been possible to work two of these with a favourable wind.

Winding was effected by a large six-blade fan, mounted behind the body. This brought the mill into the wind by drawing the superstructure round a gear-ring bolted to the hollow iron post. Several ladders and catwalks provided access to the striking gear, the winding assembly, and the brake, which, I believe, operated on a separate wheel.

The water-mill, latterly known as Hackett's mill, was still working in the early 1930s, but the windmill was, by that time, without sweeps and fan (Plate 198). The sweeps were taken down early in the First World War, and the fan in 1922 when an ailing inhabitant of West Ashling found the noise of its machinations disturbing.

During the early part of the last war parts of this unique structure were removed, leaving the tottering superstructure to stand until the late 1950s. All that remains today is the decaying water-mill building.

It is worth noting here that West Ashling water-mill, built in 1825 for corn milling, was subsequently used for making paper pulp from rags between 1832 and 1850. One of six in the county to do so, this returned to flour and provender working for the remaining 80 years of so. **[SU 808–075]**

WIND WHEELS

Although not quite within the context of this work, mentioned below are several examples of hybrid wind-wheels erected specifically for driving machinery. There were, of course, numerous wind-pumps of the so-called 'American' type to be seen at one time. These all-metal structures were used on individual farms and private houses, but have been largely superseded by the more efficient use of water-towers. These are filled by electric pumps to provide the necessary head of water.

ANGMERING

Preston Place Farm in East Preston once carried an unusual wind wheel atop the barn roofs. Erected in 1853 by the Warren family, who were local landowners,[1] this was, in fact, nearer to a small multi-sweep windmill than a true wind-wheel. Plate 199 shows this interesting structure as it remained, somewhat battered, in 1931. The eight sweeps were noted by H. E. S. Simmons as being 8ft. long by 2ft. wide, and said to have been commons. I think it more logical, and the photograph bears me out, that these were boarded and controlled by feathering gear which pivotted them, from the position show, to the 'set for wind' position required. A windshaft, or spindle in this case, was supported by a diminutuve cap which in turn was held aloft by a four-sided shingled tower. At the base of this, access was provided for reefing. Winding was carried out by a conventional fantail. The fan itself had gone by the time the photograph was taken. I gather that the striking gear and brake were both applied by ratchet and chain. The upright shaft drove line shafting in the barn below, providing power in turn for an oat-crusher, chaff-cutter, and a small water pump.

Angmering wind-wheel worked successfully for approximately 70 years[2] and was quite an attraction in the neighbourhood. By the time the Rev. P. Hemming photographed this in 1935,[2] the sweeps and fanstage had gone, and I believe that the machinery was stripped out during the early part of the last war. The tower was still standing in 1958.[3] Today, the site of East Preston Farm is entirely built over.

HEATHFIELD

Neve Bros., the local millwrights, erected an unusual wind wheel at a site in Alexander Road. I believe that this was used to power a workshop below for a Mr. Rummery.[1] The illustration used comes from the collection of the late E. Lancaster Burne (Plate 200) as do a number of others already acknowledged. This shows that this was a most interesting wind wheel erected on a smock-mill structure and was working here *c.* 1910. As may be seen, 16 tapered metal blades were pivoted between two supporting rings. These were set by a mechanism similar to patent striking gear. The open cap was winded by an eight-blade fantail. Although nothing remained of the wind wheel in recent times, the workshop itself stood until about four years ago.

PUNNETT'S TOWN

At the centre of the village of Punnett's Town, Mr. Cornford's grocery stores marks the site of a wind wheel used for powering a small carpenter's shop.

Mr. Samson Punnett was registered here in 1909.[1] The wind wheel, again erected by Neves,[2] was put up at about this time. Unlike Angmering, this was a true wind wheel consisting of an annulus of 70 fixed vanes approximately 12ft. overall diameter. The centre, approximately 7ft. diameter, was partially occupied by 14 additional blades. Winding was by a fan set behind a small smock. The whole was mounted on a turntable winding stage set above the roof of a two-storey building (Plate 201). Mr. J. Cornford was the owner at that time, and his son, who had the shop in 1977, was able to tell me that Mr. Punnett was a village carpenter and used the power for small saws and lathes, etc.

Unfortunately, there is nothing of the wheel or Mr. Punnett's workshop, for this was dismantled in 1916 as it had become derelict and dangerous. The building was subsequently converted into the single-storey shop which still stands. [TQ 623-203]

APPENDIX III

WINDMILL POPULATION

Whilst being an evident part of the Sussex landscape for over 800 years, actual windmill population by date is difficult to ascertain with accuracy. Map totals on their own are misleading, often requiring careful research and correction.[1] However, as a guide for comparison, the following note and lists are included.

During the early part of the 18th century, approximately 80 windmills were being worked and almost certainly all post-mills, smock- and tower-mills were introduced to the county during the latter part as the windmill population increased dramatically. When the first Ordnance Survey was being carried out during the early 1800s,[2] the total had risen to over 170 and increased to approximately 190 by 1823. This was not much below the peak total.

The decline in natural power-mills became evident after the 1840s. Steam plant replaced many existing mills and was incorporated in Town mills which made use of the newly-arrived railways. From then on, the Industrial Revolution began to take hold. The total standing at the end of the 19th century was something over 110, a high proportion of which were assisted by steam working.

Sadly, the 1914-18 World War had a disastrous effect. Reduced national circumstances and priorities put the seal of fate on a great many mills, whose future had, in any case, been in doubt, leaving 60 to 70 in some form or other. These were subject to demolition, auxiliary power working, and in a few early cases actual preservation. Just prior to the outbreak of the Second World War in 1939, there were eight working windmills of a total of 45 or so standing.

The list on page 182 shows what is left of the past hundreds; 89 extant windmill structures and remains. Of the 28 shown in the first section, none are active, but three can still work– one post-mill, one smock-mill, and one tower-mill.

Following is a list of windmills known to have stood in the last 170 years and shows those mentioned in this work.

APPENDIX IV

LIST BY PARISH OF WINDMILLS STANDING TODAY AND BYGONE MILLS KNOWN TO HAVE STOOD DURING THE LAST 170 YEARS

S; Standing today. R; Remain only. OR: Obscure remains. G: Gone. M: Mentioned in text

Aldingbourne	two mills	G
Alfriston	tower-mill	RM
Amberley	Old mill	G
Angmering	Jerusalem mill	G
Angmering	Luck's post-mill	RM
Angmering	tower-mill	RM
Appledram	Dell Quay, post-mill	RM
Arlington	Windover, post-mill	OR, M
Arundel	cement mill	G
Arundel	Atfield's mill	G
Arundel	South Marsh tower-mill	SM
Ashington	Old mill	G
Ashurst	post-mill	RM
Baldslow	post-mill	GM
Baldslow	smock-mill	RM
Barcombe	post-mill	G
Barnham	Feaver's mill	G
Barnham	Old mill	GM
Barnham	tower-mill	SM
Battle	*King's Head* post-mill	GM
Battle	smock-mill	SM
Battle	Telham Hill post-mill	G
Beckley	Old mill	G
Beckley	New mill (removed to Petham, Kent)	G
Beddingham	Old mill	G
Beeding (Upper)	post-mill	G
Berwick	post-mill	G
Bexhill	Mount Idle mill	G
Bexhill	Black mill	G
Bexhill	Sidley smock-mill (moved to Leigh in Kent 1928)	OR, M
Bexhill	Down post-mill	RM
Billingshurst	Sprinks mill	GM
Billingshurst	Hammond's smock-mill	RM
Birdham	post-mill	G
Birdham	pumping-mill (smock)	GM
Bishopstone	Hoisting smock-mill at Tide mills	GM
Blatchington (East)	Black mill	G
Blatchington (East)	pumping tower-mill	GM
Blatchington (West)	smock-mill	SM
Bolney	smock-mill	G

Bosham	Old mill	G
Boxgrove	Halnaker tower-mill	SM
Brede	post-mill	G
Brightling	Beacon mill	G
Brighton	Black Rock	G
Brighton	Taylor's post-mill	G
Brighton	Clifton Gardens	G
Brighton	Vine's post-mill	G
Brighton	Park mill	G
Brighton	Toronto Terrace	G
Brighton	Windmill Street (two)	G
Brighton	Murrell's post mill	G
Brighton	Preston mill	G
Brighton	Hanover mill	G
Brighton	Lashmar's Old mill	G
Brighton	Lashmar's New mill	GM
Brighton	Port Hall mill	G
Brighton	Bear post-mill	G
Brighton	Streeter's mill	G
Brighton	Rosehill tower-mill (Cuttress's)	GM
Brighton	Hodson's black smock-mill	GM
Burgess Hill	post-mill	G
Burwash	post-mill	G
Burwash	Witherenden	G
Burwash	Rockhill–Weald post-mill	GM
Catsfield mill	post mill	G
Chailey	post-mill	GM
Chailey	Heritage smock-mill	SM
Chailey	South Common smock-mill	RM
Chichester	St. James's mill	G
Chichester	Broyle mill (two)	G
Chichester	Portfield mill	G
Chiddingly (The Dicker)	Willard's mill	GM
Chiddingly (The Dicker)	Black mill	G
Chiddingly (The Dicker)	Golden Cross post-mill	RM
Chiddingly (The Dicker)	Lower Dicker post-mill	RM
Chidham mill	old mill	G
Chiltington (West)	Meeten's smock-mill	SM
Clayton	Dungate post-mill	RM
Clayton	(Jack) tower-mill	SM
Clayton	(Jill) post-mill	SM
Climping	smock-mill	RM
Cold Waltham	Watersfield Common	G
Cold Waltham	Watersfield smock-mill	GM
Compton	old mill	G
Crowborough	Beacon post-mill	OR, M
Crowborough	Pratt's tower-mill	RM
Cuckfield	Beech Farm smock-mill	RM
Cuckfield	Whiteman's Green post-mill	GM
Cuckfield	Kennard's mill	G
Dallington	smock-mill	RM
Danehill	New mill	G
West Dean	Binderton Rook mill	G
Ditchling	Elphick's Farm	RM
Durrington	High Salvington post-mill	SM
Earnley	Sommerly smock-mill	SM

Eastbourne	Manor Hill	G
Eastbourne	Parsonage (white) mill (Watt's Lane smock-mill)	GM
Eastbourne	Black mill	G
Eastbourne	Rodmill Farm	G
Eastbourne	Ocklynge tower-mill	GM
Eastbourne	Hurst's or St. John's mill	RM
Ewhurst	Brasses' mill	G
Ewhurst	Beacon mill	G
Ewhurst	Staplecross smock-mill	RM
Fairlight	old mill	G
Fairlight	Batchelor's pump post-mill	G
Falmer	old mill	G
Falmer	Race Hill post-mill	G
Felpham	Black mill	G
Felpham	White smock-mill	G
Ferring	John Olliver's post-mill	GM
Findon	old mill	G
Firle, West	two post-mills	G
Fishbourne	post-mill	OR, M
Fletching	old mill	G
Forest Row	old mill	G
Framfield	Mount Ephraim post-mill	GM
Framfield	Blackboys post-mill	OR, M
Frant	two	G
Friston	old mill	G
Friston	post-mill	G
Funtington (West Ashling)	hollow post-mill at water-mill	OR, M
Glynde	post-mill	GM
Glynde	hollow post pumping-mill	OR, M
East Grinstead	North End mill	G
East Grinstead	Ashurst Wood	G
West Grinstead	Littleworth (Jolesfield)	RM
Guestling	old mill	G
Guestling	smock-mill	GM
Hailsham	Upper post-mill	GM
Hailsham	Lower—Hamlin smock-mill	GM
Hailsham	Harebeating post-mill	RM
Hamsey	post-mill	G
Hamsey	Offham smock-mill (Racehill)	G
Hastings	Summerfield's mill	G
Hastings	Croft Road French's smock-mill	OR, M
Hastings	Priory Road mills (five)	G
Hastings	Town mill	G
Heathfield	Sandy Cross post-mill	G
Heathfield	Broad Oak post-mill	G
Heathfield	Mutton Hall	OR, M
Heathfield	Cade Street	G
Heathfield	Punnett's Town post-mill	GM
Heathfield	Punnett's Town smock-mill	SM
Heathfield	Punnett's Town Saw smock-mill	GM
Hellingly	old mill	G
Henfield	Old post-mill	G
Henfield	New post-mill	GM
Highbrook	Hammingden smock-mill	GM
East Hoathly	post-mill	G

West Hoathly	Selsfield Common smock-mill	G
Hooe Common	post mill	G
Hooe Common	smock-mill	G
Horeham	smock-mill	GM
Horsham	Wimblehurst mill	G
Horsham	Champion's mill	G
Horsham	Warnham mill	G
Horsham	Compton mill	G
Horsham	Roffey mill	G
Horsham	Southwater—Cripplegate smock-mill	G
Little Horsted	old mill	G
Hunston	smock-mill	RM
Hurstmonceux	Windmill Hill early post-mill	G
Hurstmonceux	Windmill Hill post-mill	SM
Hurstmonceux	Bodle Street post-mill	G
Hurstmonceux	Cowbeech smock-mill	RM
Hurstpierpoint	Peskett's post-mill	G
Icklesham	Telegraph smock-mill (Broad Street)	G
Icklesham	Hog Hill mill	SM
Iden	Post-mill	G
Ifield	post-mill	OR, M
Keymer	Oldland post-mill	SM
Kingston	smock-mill	RM
Lamberhurst	Sussex mill	G
Lancing	post-mill	G
Lewes	Ashcombe six-sweep smock-mill	GM
Lewes	Town smock-mill	RM
Lewes	Shelly's smock-mill	RM
Lewes	Southern post-mill	OR, M
Lewes	Spital post-mill	OR, M
Lewes	Malling post-mill	RM
Littlehampton	Cudlow post-mill	G
Littlehampton	Arun tower-mill	GM
Lowfield Heath (belonged to Surrey until April 1974)	post-mill	SM
Lyminster	Brookfield	G
Lyminster	Toddington mill	G
Mark Cross	Walter's tower-mill	RM
Mayfield	Luggers Crouch mill	G
Mayfield	Argos Hill mill (post)	SM
Newenden	pumping-mill	G
Newhaven	Bolen's mill	GM
Newhaven	smock-mill	GM
Newick	old mill	G
Ninfield	post-mill	G
Ninfield	Lunsford Cross thorne mill (smock)	RM
North Chapel	old mill	G
Northiam	High Park post-mill	RM
Northiam	old mill	G
Northiam	Millcorner mill	G
Northiam	Millcorner mill	G
Nutley	post-mill	SM
Ore	'Cheale's' white smock-mill	RM
Ore	'Down' black smock-mill	RM
Oving	old mill	G

Pagham	post-mill	G
Pagham	Nyetimber tower-mill	SM
Patcham	Ballard's mill (smock)	G
Patcham	Waterhall mill (tower)	SM
Peasemarsh	smock-mill	G
Petworth	old mill	G
Plaistow	Pullen's mill	G
Playden	black mill	G
Playden	old mill	G
Playden	Playden mill	G
Playden	smock-mill	G
Playden	old post-mill	G
Portslade	East Hill post-mill	G
Portslade	Copperas Gap mill	G
Pulborough	Heath mill	G
Pulborough	Pulborough mill	G
Pulborough	Nutbourne tower-mill	SM
Ringmer	Glyndebourne post-mill	RM
Ringmer	Broyle post-mill	G
Ripe	old mill	G
Rodmell	old mill (post)	G
Rottingdean	post mill	GM
Rottingdean	Beacon smock-mill	SM
Rudgwick	Rudgwick mill	G
Rudgwick	Honeylane mill	G
Rusper	old mill	G
Rustington	Bridge mill	G
Rustington	Sea Field mill	G
Rustington	Rustington mill	G
Rye	Wilkinson's mill	G
Rye	old post-mill	G
Rye	Barry's smock-mill	GM
Rye	Barry's smock-mill replica	SM
St. Leonard's	Tivoli old mill	GM
St. Leonard's	Tivoli new mill	GM
St. Leonard's	Draper's mill (Silverhill smock-mill)	GM
Salehurst	Silverhill mill	G
Seaford	Sutton mill	GM
Selham	Moorland Farm	G
Selsey	Medmerry tower-mill	SM
Shipley	Honeypoles mill	G
Shipley	King's smock-mill	SM
Shipley	Coolham post-mill	G
Old Shoreham	post-mill	GM
New Shoreham	old mill	OR, M
Sompting	pumping-mill	G
Southwick	Southwick mill	G
Southwick	cement mill	G
Steyning	post-mill	G
North Stoke	post-mill	G
Storrington	Botting's mill	G
Storrington	Storrington mill	G
Sullington	post-mill	RM
Twineham	Hooker's mill (hollow post-mill at water-mill)	OR, M
Uckfield	old mill	G

Udimore	post-mill	GM
Upmarden	Haslet mill	G
Upmarden	Upmarden mill (apple down)	G
Wadhurst	Cousley Wood mill	G
Wadhurst	Riseden post-mill	RM
Wadhurst	Butcher's Wood mill	G
Walberton	Avisford mill	G
Walberton	Short's smock-mill	OR, M
Waldron	Cross in Hand old post-mill	RM
Waldron	Cross in Hand new post-mill	SM
Warbleton	Summer Hill post-mill	OR, M
Warbleton	Chapman's Town smock-mill	RM
Wartling	Wartling mill	G
Wartling	Boreham Street post-mill	G
Washington	Rock Common smock-mill	RM
Westfield	Westfield church	G
Westfield	White mill	G
Westfield	North mill	G
Westham	Westham mill	G
Westham	Hollow post pumping-mill; reconstructed at Singleton open air museum	G/OR, M
Westham	Stone Cross tower-mill	SM
Whatlington mill	old mill	G
Willingdon	old mill	G
Willingdon	Polegate tower-mill	SM
Winchelsea	St. Leonard's mill (post)	RM
Wisborough Green	Amblehurst mill	G
Wisborough Green	Champion's mill (smock)	RM
Wittering (East)	tower-mill	RM
Worth	Effingham mill	G
Worth	Crawley Down	G
Worthing	Heene post-mill	
Worthing	Cross Street and Isteds post-mill	GM
Worthing	Sea post-mill	G
Worthing	Navarino North smock-mill	GM
Worthing	Navarino South tower-mill	GM
Worthing	Cissbury mill (post)	G
Worthing	pumping-mill	G

LARGE WIND-WHEEL STRUCTURES

Angmering	Preston Place Farm	GM
Eastbourne	Bullock Down	G
Heathfield	Alexandra Road	GM
Heathfield	Punnett's Town	OR, M

APPENDIX V

LIST OF WINDMILLS AND REMAINS EXTANT IN SUSSEX, 1979

Mills Today

Post	Smock	Tower
Clayton	Battle	Arundel
Cross in Hand	Chailey	Barnham

Mills Today (*continued*)

High Salvington	Earnley	Clayton
Hurstmonceux	Punnett's Town	Halnaker
Icklesham	Rottingdean	Nutbourne
Keymer	Rye	Nyetimber
Lowfield Heath	Shipley	Patcham
Mayfield	West Blatchington	Polegate
Nutley	West Chiltington	Selsey
		Stone Cross

Notable Remains

Post	Smock	Tower
Angmering	Baldslow	Alfriston
Appledram	Billingshurst	Angmering
Ashurst	Chailey South	Crowborough
Bexhill Down	Climping	East Wittering
Clayton (Duncton)	Cowbeech	Eastbourne (Hurst's mill)
Cross in Hand (old mill)	Cuckfield	Mark Cross
Dicker (Golden Cross)	Dallington	
Dicker (Lower)	Hunston	
Harebeating	Kingston	
Lewes (Malling)	Lewes (Shelly's)	
Northiam	Lewes (Town)	
Ringmer	Littleworth	
Riseden	Lunsford's Cross	
Sullington	Ore (black mill)	
Winchelsea	Ore (white mill)	
	Staple Cross	
	Warbleton (Chapman's Town)	
	Washington	
	Wisborough Green	

Hybrids

Horizontal mill at Pashley Down, Eastbourne	Utility mill at Ditchling, Elphick's Farm

Obscure Remains

Post	Smock	Tower
Arlington (Windover)	Bexhill (Pankhurst's mill)	
Blackboys	Hastings, Croft Road	
Crowborough	Walberton	
Fishbourne		
Glynde (hollow post-mill)		
Heathfield (Mutton Hall)		
Ifield		
Keymer (model windmill)		
Lewes (Southern)		

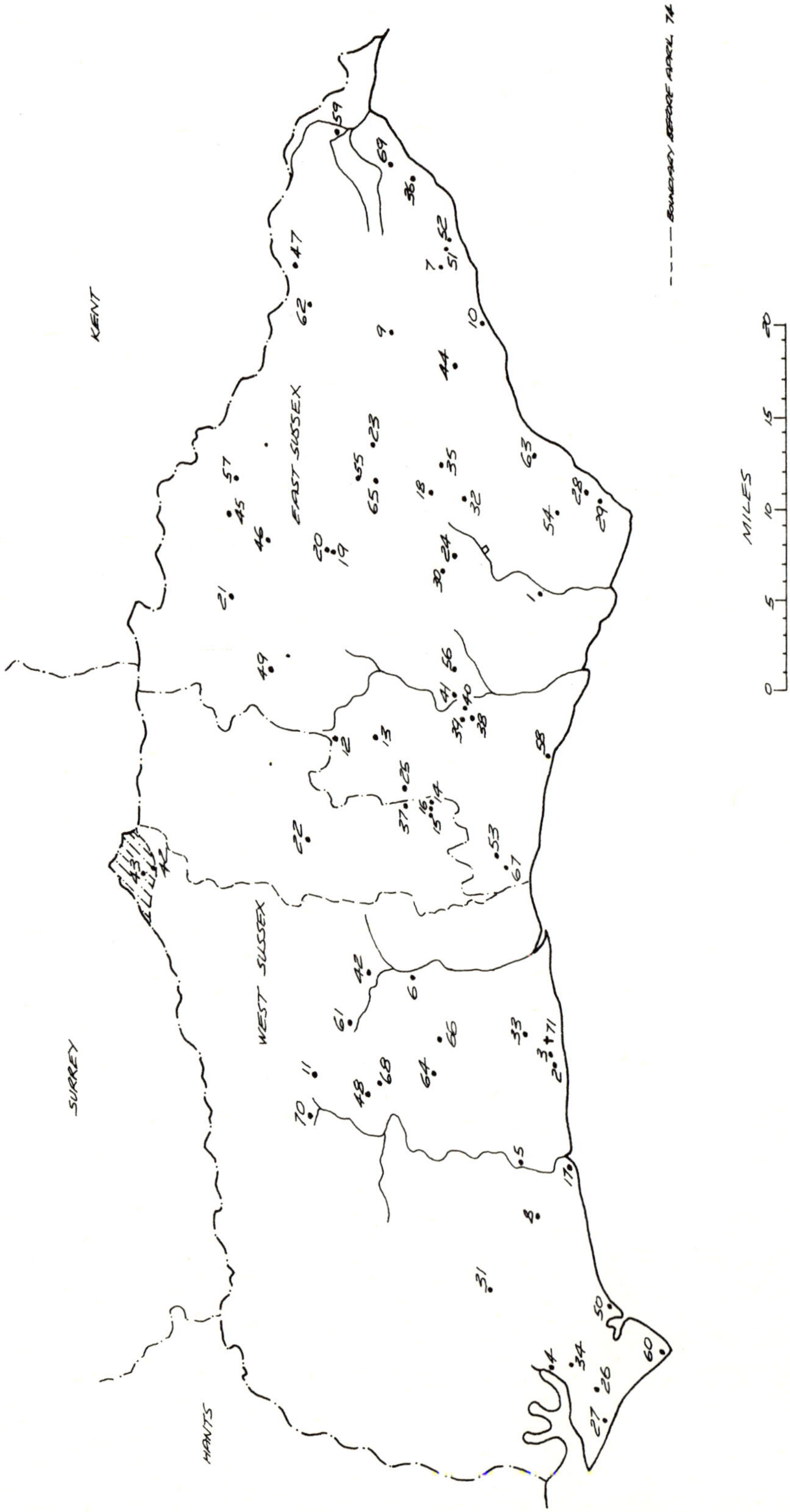
KENT
SURREY
HANTS
EAST SUSSEX
WEST SUSSEX
MILES
0
5
10
15
20

Fig. XI. Key to notable windmill structures and remains extant in Sussex (1979).

Obscure Remains (*continued*)

Lewes (Spital)
Shoreham (Good Friday mill)
Singleton (hollow post-mill)
Twineham (hollow post-mill)
Warbleton (Summer mill)
West Ashling (hollow post mill)

Wind-wheel

Punnett's Town

BIBLIOGRAPHY AND REFERENCE

Batten, M. I., *English Windmills,* The Architectural Press. 1930.
Bennett and Elton, *History of Corn Milling,* Vol. II, Simpkin Marshall. 1899.
Betjeman and Gray, *Victorian and Edwardian Brighton,* Batsford. 1973.
Charnock, G. F., *Mechanical Technology,* Constable. 1915.
Freese, S., *Windmills and Millwrighting,* David and Charles. 1971.
Fowel and Foord Hughes, *Windmills in Sussex,* Walker's *Quarterly.* 1930.
Hemming, P., *Windmills in Sussex,* Daniels. 1936.
Hopkins, R. T., *Old English Mills and Inns,* Palmer. 1927.
Hopkins, R. T., *Old Watermills and Windmills,* Allen. 1933.
Lockwood, Sir Joseph, *Flour Milling,* Henry Simon. 1972.
Lucas, E. V., *Highways and Byeways of Sussex,* Macmillan. 1903.
Mais, S. P. B., *Sussex,* Methuen. 1950.
Mais, S. P. B., *England of Windmills,* Dent. 1933.
Marten, E. A., *Life in a Sussex Windmill,* Allen. 1921.
Mee, A., *Sussex,* Hodder and Stoughton. 1971.
Wailes, R., *The English Windmill,* Routledge. 1954.
Wailes. R., *Windmills in England,* Skilton. 1976.

SOUVENIR GUIDES AND BOOKLETS, ETC.

Nutley Windmill Restoration, Nutley Windmill Appeal. 1974.
Polegate Windmill, Eastbourne and District Preservation Trust. 1967.
Rottingdean Windmill, The Rottingdean Preservation Society. 1977.
Salvington Mill, Worthing Museum. 1976.
Shipley Mill, West Sussex County Council. 1966.
Windmills and Argos Hill (Mayfield), Wealden District Countil. 1975.
Various articles and news items in *Sussex County Magazine.* 1927–1956.

RESEARCH SOURCES CONSULTED

The H. E. S. Simmons collection ('Sussex Windmills'), Brighton Reference Library.
The J. M. Baines collection, Hastings Reference Library.
The Gurney Wilson collection, Hove Central Library.

The 'Sussex Windmill' collection, Worthing Reference Library.
East Sussex Records Office, Lewes.
West Sussex Records Office, Chichester.
Sussex Archaeological collection, Barbican House, Lewes.

SUSSEX MAPS CONSULTED

Budgen: One inch map of Sussex. 1724.
Yeakell and Gardner: Two inch map of Sussex. 1778/85.
Gardner Yeakell/Gream: One inch map of Sussex. 1795.
Mudge (early printing): First one inch Ordnance Survey. 1813.
Greenwood: One inch map of Sussex. 1823/24.
Tithe Award Survey: Approx. 25 inches to the mile. 1838/53.
Ordnance Surveys: First 6 inch to the mile. 1873/76.
Ordnance Surveys: Modern one inch and 1:50,000.

SUSSEX DIRECTORY REFERENCES (Author and H. E. S. Simmons)

From Kelly and isolated editions of Melville and Pigot.

AUTHOR'S REFERENCE NOTES

Introduction

1. S.A.C., xxix, p. 143.
2. S.A.C, v, p. 274.

Page 14

1. *Dictionary of Sussex Dialect* (Augmented), Parish, 1875/1957.

Page 33

1. Various estimates put this at between 1,500ft. and 3,000ft.

Jill, Clayton, page 36

1. S.A.C., 821.
2. Brighton Reference Library. S.B.912, 1850, is the last to show her. S.B. 912, 1853, and onwards do not.
3. Survey date of First 6-inch Ordnance Survey.
4. F.G.S., Bramwell in Musgrave's *Brighton*, p. 221.
5. K., 1870, shows Mitchell. Hammond's patent taken out in 1873.
6. K, 1903.
7. Mr. C. Hoad.
8. Letter from E. A. Martin, *Sussex County Magazine*, Vol. X, p. 219.
9. Mr. F. W. Gregory, the noted Sussex windmillian, believes this to be either chestnut or oak.

Cross in Hand, page 38

1. E.S.R.O., TD/E6.
2. Memoir of Eli Ashdown, Farncombe, 1904.
3. *Sussex Weekly Advertiser*, 17 July 1809 Notice.
4. Mr. C. J. Newnham.
5. Mr. S. Ashdown.
6. Neve's account book, 1921–30.

High Salvington, page 41

1. Said to be in 1774 by Sun Fire Insurance, but the author has been unable to substantiate this.
2. Worthing museum, No. 12.

Hurstmonceux, page 43

1. *Sussex Weekly Advertiser*, 30 August 1819, and *Sussex Advertiser*, 20 January 1823 notices.
2. K, 1845, 1878.
3. H.E.S.S., from the Harmer family.
4. K., 1895.
5. C. E. Hammond's patent No. 1654, 1873. This shows the arrangement as applied at Clayton tower-mill and is identical in principle to H.X.
6. Boaz Medhurst took over from Samuel, *c.* 1870.

Icklesham, page 47

1. 31 December 1781, John Skinner, miller.
2. John Skinner, proprietor.
3. H.E.S.S.
4. Pigot, 1834.
5. Kelly.
6. Kelly and H.E.S.S., from Gardner family.

Keymer, page 48

1. *Walker's Quarterly*, Nos. 29 and 30, 1930.
2. H.E.S.S., from F. Harrison, M.A.
3. E.S.R.O.
4. Melville.
5. Kelly.
6. Mr. R. J. de Little.
7. H.E.S.S., from Mr. D. Driver.

Lowfield Heath, page 50

1. H.E.S.S. and *Windmills of Surrey and Inner London*, Farries and Mason.
2. Gurney Wilson collection.
3. K., 1862.

Mayfield, page 52

1. H.E.S.S., from *Rotherfield*, by Catherine Pullein.
2. Windmills and Argos Hill booklet, Wealden District Council.
3. Kelly and Melville.
4. E.S.R.O. TD/E133.
5. Neve's account book, 1921–30.

Nutley, page 54

1. Gurney Wilson collection, however, only an obscure mark dated 1820 remains today.
2. E.S.R.O., TD/E151.
3. Kelly.
4. Nutley Windmill Restoration—Uckfield and District Preservation Society.

Battle, page 56

1. H.E.S.S., Mr. A. N. Neve.
2. It is traditionally held that this should be King's Mead, as there is strong evidence that Caldbec Hill was used by King Harold as a mustering point before the Battle of Hastings. See also *The Enigma of Hastings*, by Edwin Tetlow, 1974, pp. 151–153 in which he tells us that the mill marks the site of a Hoar apple tree, mentioned in the *Anglo-Saxon Chronicle* as the rallying point.
3. Date stone A.P. 1840 in granary wall.
4. Kelly.

Chailey, page 57

1. *Official Heritage History* (1948).
2. *Sussex County Magazine,* Vol. VI, p. 822, from Mr. Godley, Mr. Beard's grinder at Chailey smock-mill.
3. E.S.R.O., TD/E41.
4. *Sussex Advertiser,* 26 August 1845 and 29 June 1864.
5. Kelly.

Earnley, page 59

1. H.E.S.S., E.S.R.O. Defence schedule, 1801-3.
2. K., 1845.
3. From Mr. C. N. Darby.

Punnett's Town, page 60

1. Also Neves account book, 1921-30.
2. *Sussex Advertiser,* 7 September 1859.
3. Mr. A. D. Dallaway.
4. *Walker's Quarterly,* Nos. 29 and 30, 1930.

Rottingdean, page 63

1. According to official history of the mill T B 1802 is to be found inscribed inside the mill.
2. H.E.S.S. and Official history.
3. Story of Rottingdeam Windmill, Rottingdean Preservation Society.
4. Work carried out between 1960 and 1976 at a total cost of £10,442.

Rye, page 65

1. H.E.S.S.
2. E.S.R.O., TD/E1.
3. *Sussex Advertiser,* 16 September 1879, sale notice.
4. Kelly.
5. Mr. Alan Webb.
6. H.E.S.S., from Neves.

Shipley, page 66

1. Kelly.
2. Official guide: Shipley Mill, West Sussex County Council, from the Jebb family.
3. H.E.S.S.
4. Powell family.

West Blatchington, page 68

1. Mr. Ray, Borough of Hove Planning Department.
2. *Brighton and Hove Herald/Evening Argus.*
3. The author's opinion is that the mill was erected *c.* 1820.
4. Hove Library print 2685.
5. E.S.R.O.
6. H.E.S.S.
7. *Old Windmills and Watermills,* by R. Thurston Hopkins, Alan, 1933, p. 198.
8. Kelly.

West Chiltington, page 71

1. *West Chiltington,* by Mrs. Saunders Jacobs.
2. W.S.R.O., TD/W159.
3. This indicates that this was not an old mill re-erected M B.
4. Mrs. H. B. Roscow.
5. Kelly.
6. H.E.S.S., from Mr. Powell.

Arundel, page 72

1. *Sussex Advertiser* 10/9/1832 sale notice: 19 years unexpired on 21-year lease.
2. W.S.R.O., TD/W5.
3. H.E.S.S., from the Bartlett family.

Barnham, page 73

1. Kelly.
2. Mr. R. C. Reynolds.
3. H.E.S.S.
4. *Brighton Herald,* 23 January 1830.
5. Photographs, S.A.C., Mr. Beasley.

Clayton, page 76

1. Gurney Wilson Collection, also; Figg's map, 1861, shows Duncton only, and First 6-inch Ordnance Survey shows Jack and Jill and roundhouse of Duncton in 1873.
2. Kelly shows J. Mitchell till 1870.
3. H.E.S.S., from Gurney Wilson.
4. Charles Edwin Hammond, Clayton mills, Patent No. 1654, 1873.
5. Letter from E. A. Martin, *County Magazine,* Vol. X, p. 219.
6. Allen and Donaldson, 1921.
7. Mr. Henry Longhurst, also *Jack by the Hedge,* by Nancy Price, 1942.
8. Mr. E. Hole.
9. Shown in Kelly, 1874.

Halnaker, page 78

1. W.S.R.O., Goodwood Estate Archives, Vol. II, E287,
2. Ditto, E32.
3. Ditto. Settlements and Trust, 1087–89.
4. H.E.S.S., from Mr. Watkins, also; Boxgrove tithe map, W.S.R.O., TD/W19, 1839, shows Duke of Richmond as owner and Charles Adam at Mill Cottage.
5. Painting by Turner (early 19th century). See *The English Windmill,* by Rex Wailes, 1954, Platexix.
6. Mr. A. D. Dallaway.
7. Kelly.

Nutbourne, page 80

1. H.E.S.S.
2. H.E.S.S., from Mr. Meeten of West Chiltington.
3. S.A.C., 821.
4. H.E.S.S., from Mr. Stillwell and Kelly.

Nyetimber, page 81

1. Windmill collection, Worthing Reference Library.
2. William Adams first shown in Kelly, 1845.
3. H.E.S.S.,
4. *The Miller* sale notice, 16 August 1897.
5. Kelly.
6. H.E.S.S., from Mr. Durman.

Patcham, page 82

1. H.E.S.S., from the Harris family.
2. H.E.S.S., from Mr. Bull, the last miller at Patcham.
3. Information from property deeds kindly supplied by Mr. C. J. Dowd.
4. Mr. F. Harris, *Evening Argus.*

Polegate, page 85

I am indebted to the following for the details of the history and restoration:

1. Miss Vera Hodsoll.
2. Mr. Lawrence Stevens.
3. Mr. F. W. Gregory

4. Mrs. Bertha Terry, daughter of Albert Ovenden.
5. 'Polegate Windmill', Eastbourne and District Preservation Trust, Ltd.

Selsey, page 87
1. H.E.S.S.
2. *Brighton Herald.*
3. Melville and Kelly.
4. H.E.S.S., from Bob Barber, one of Holloway's men.

Stonecross, page 89
1. From Mr. R. Hall and H.E.S.S., from H. Dallaway and Neves.
2. Kelly's.

Angmering, page 92
1. Melville and Kelly.
2. H.E.S.S., from the Luck family.
3. Tithe map, W.S.R.O., TD/W35.

Appledram, page 93
1. H.E.S.S.
2. 23 February, 1807
3. 23 July 1810.
4. Tithe map 1838, W.S.R.O., TD/W4.
5. Melville and Kelly.

Ashurst, page 94
1. Gurney Wilson collection.
2. Melville and Kelly.
3. *English Windmills,* Vol. I, M. I. Batten, 1930, p. 101.

Bexhill Down, page 95
1. A full account of the family association with the mill from Mr. C. Hoad, 1975/76.
2. H.E.S.S., from Mr. John Hoad.
3. E.S.R.O., tithe map TD/E141, 1839.

Clayton (Duncton), page 96
1. Mr. C. Hoad.
2. Gurney Wilson collection. Note of indenture 99-year lease from 2 September 1765.
3. 22 October 1787, 20 November 1809, 16 April 1810, 16 September 1816.
4. TD/E72

Cross in Hand (Old), page 98
1. From Mr. C. J. Newnham.
2. H.E.S.S.
3. Memoir of Eli Ashdown, Farncombe, 1904.
4. From Mr. S. Ashdown.

Dicker (G.C.), page 99
1. Melville and Kelly.
2. H.E.S.S., from the Wickens family.
3. Mr. S. Ashdown.
4. E.S.R.O., TD/E105.
5. Kelly.
6. *The English Windmill,* by Rex Wailes, p. 46.

Dicker New Mill, page 99
1. H.E.S.S.
2. Tithe map, E.S.R.O, TD/E105.
3. Kelly.
4. Mr. S. Ashdown.

5. H.E.S.S.,. from the Ovenden family.
6. H.E.S.S., from the Gurney Wilson collection.

Harebeating, page 100

1. *Sussex Advertiser,* 21 December 1869.
2. Kelly.
3. H.E.S.S.
4. H.E.S.S., and Mr. F. W. Gregory.

Lewes (Ashcombe) (Kingston), page 101

1. Gurney Wilson Collection.
2. 'The Old Windmills of Lewes', article by H. E. S. Simmons, Sussex County Magazine, Vol. XV, p. 345.
3. Kelly.
4. H.E.S.S. from Mr. John Ashdown.
5. Photograph by F. W. Gregory in Mill notes, No. 6, S.P.A.B.
6. Letter from Sussex Archaeological Society, 14 December 1975.

Lewes (Malling), page 103

1. H.E.S S.
2. S.A.C., Vol. 73, p. 207, notes regarding South Malling mill deeds.
3. H.E.S.S., writing in *Sussex County Magazine,* Vol. XV, p. 352.
4. *Sussex Weekly Advertiser,* 30 April 1804.
5. Ditto, 12 December 1821.
6. Kelly.

Northiam, page 104

1. H.E.S.S.
2. Letter from Mrs. Perigoe (Mr. C. J. Banister's daughter),
3. Tithe map, E.S.R.O., TD/E96.
4. Kelly.
5. H.E.S.S., from Mr. A. A. Goble.

Ringmer, page 105

1. *Brighton Herald,* 13 June 1925, also there are two photographs of the collapsed remains in Victoria and Edwardian Windmills and Watermills, J. K. Major and M. Watts, Batsford, 1977.
2. H.E.S.S., from the Holter family.
3. Tithe map, E.S.R.O., TD/E137.

Riseden, page 106

1. E.S.R.O.
2. Tithe map, E.S.R O., Wadhurst.
3. Kelly.
4. H.E.S.S.
5. H.E.S.S., from Neves.

Sullington, page 107

1. H.E.S.S.
2. Kelly.
3. H.E.S.S., from Gurney Wilson.

Winchelsea, page 108

1. In Winchelsea town museum there is a painting on wood which shows 'St. Leonard of Winchelsea A.D. 1300'.
2. Winchelsea town museum, 'History of the Mill'.
3. E.S.R.O., TD/E90.
4. Shown at the old site on the First Ordnance Survey map, surveyed *c.* 1808. Shown at the present site on Greenwood's map, surveyed 1823. Built at old site in 1760 and either moved to or rebuilt on present site between 1808 and 1823. M.B.

5. *Sussex Advertiser.*
6. Mr. E. Hole.
7. H.E.S.S., Melville and Kelly, John sharp's last shown in 1895.

Baldslow, page 109

1. From Mr. William Draper of Silver Hill mill, St. Leonard's, S.C.M., 1930, p. 1064.
2. J. M. Baines collection, Hastings Reference Library–*Hastings and St. Leonard's News,* 6 July 1855.
3. Deeds in the possession of Mrs. J. Greenall.
4. *Hastings and St. Leonard's Pictorial Advertiser,* 6 April 1916.
5. Letter from Miss M. Foster, S.C.M., 1930, p. 992.

Billingshurst, page 110

1. H.E.S.S.
2. Pigot.
3. W.S.R.O. Tithe map, TD/W13.
4. Kelly.
5. H.E.S.S., from Mr. Powell, miller at Coolham and Shipley.
6. Mr. C. F. Pierce.
7. Gurney Wilson Collection.

Chailey South, page 111

1. H.E.S.S., from Mr. Avery.
2. Kelly.
3. *Sussex Advertiser,* 4 June 1862.
4. Gurney Wilson collection.
5. 'Sussex Curiosities', No. 2, a photograph in *West Sussex Gazette,* 1922, Worthing Reference Library.
6. Mrs. N. Styles.

Climping, page 112

1. A succession of mills are shown here from 1610; the date 1799 claimed by the Rev. P. Hemming (contemporary of Earnley, *c.* 1800, and Hunston, 1801, M.B.).
2. H.E.S.S.
3. Kelly.
4. Mr. F. W. Gregory.
5. Survey by Rex Wailes.
6. Mr. E. Hole.

Cowbeech, page 113

1. H.E.S.S.
2. E.S.R.O., Tithe map, TD/E89.
3. Kelly.
4. Mr. G. A. Catt.
5. 18 June 1900.

Cuckfield (Beech Farm), page 114

1. Gurney Wilson collection, Hove Library.

Dallington, page 115

1. Details of lease; one acre at Dallington Street to Mr. Albert Geering, at £6 per annum from The Ashburton Estate, 99 years from 24 September 1852.
2. Popularly attributed to Neve. Stephen Neve was, in fact, working for Upfields at the time.
3. Kelly.
4. *The Miller,* 21 April 1902.
5. A contemporary postcard, dates 1913, shows this.

Hailsham (Hamlin's), page 116

I am indebted to Mr. Gilbert Catt, who is the grandson of David, for a complete and substantial account of the windmill, power-mill, and the family working here.

Hunston, page 117

1. H.E.S.S., from Mr. C. Hodson.
2. Property deed, Mr. M. J. Miller.
3. Kelly.
4. W.S.R.O., Tithe map TD/W70.

Kingston, page 118

1. H. E. S. Simmons, 'The Old Windmills of Lewes', *Sussex County Magazine*, Vol. XV, p. 345.
2. E.S.R.O., Tithe map TD/E67.
3. H.E.S.S., from Mr. John Ashdown.
4. *Sussex Weekly Advertiser*, 20 and 27 December 1802.
5. From various rather poor photographs.
6. Kelly.

Lewes (Shelly's), page 119

1. H. E. S. Simmons, 'The Old Windmills of Lewes' *Sussex County Magazine*, Vol. XV, p. 347.
2. Gideon Mantell's Diary, February 1819, mentions 'the new mill' at the 'site of the barracks'. The barracks were last shown on the early edition of the one-inch Ordnance Survey map, 1813, and there was open ground here until the late 1840s, when the prison was built.
3. Kelly and Melville.
4. H.E.S.S., from Alderman John Every.

Littleworth, page 120

1. *Sussex Weekly Advertiser*, 26 May 1788.
2. *Walker's Quarterly*, Nos. 29 and 30, 1930.
3. Kelly.
4. As noted by Rev. P. Hemming, *Windmills in Sussex*, p. 64.

Lunsford's Cross, page 122

1. H.E.S.S.
2. Notice in *Sussex Advertiser*, 7 April 1874.

Ore (Black), page 123

1. J. M. Baines collection, from Mr. A. W. Thomas.
2. Kelly.
3. H.E.S.S.
4. From Mrs. D. K. Freeland.

Ore (White), page 123

1. A very good first-hand account was given to J. M. Baines, *Historic Hastings*, p. 270; also *Hastings and St. Leonards Observer*, 19 May 1900.
2. H.E.S.S.
3. Pigot.
4. Melville.
5. Kelly.
6. A notice in the *Sussex Advertiser*, 10 March, 1874, refers to a mill in the area as 'leasehold for a term of 99 years from 29th September 1812' would appear to refer to the white mill. M.B.

Staplecross, page 124

1. J. M. Baines, from Mr. E. M. Dadswell, great-grandson of Thomas Martin. 'After trying the mill himself, rented it out'.
2. 11 November 1816.
3. Kelly.
4. H.E.S.S., from Mr. Goble.
5. H.E.S.S., from H. R. Hawksley.

6. Full account from Mr. A. D. Dallaway. M.B.
7. *The Miller,* 4 November 1901.

Warbleton, page 125

1. *Sussex Advertiser.*
2. H.E.S.S., from Mr. Ellis.
3. First 6-inch Ordnance Survey map.
4. E.S.R.O., TD/E50.

Washington, page 126

1. Air Commodore J. F. Davis, also, not shown on Greenwood's map, 1825, date 1827, cut in brickwork.
2. *Sussex Advertiser.*
3. W.S.R.O., Tithe map TD/W137.
4. Kelly.
5. Mr. S. Ashdown.
6. H.E.S.S.

Wisborough Green, page 127

1. *Sussex Weekly Advertiser.*
2. W.S.R.O., TD/W149.
3. The old course is shown a quarter of a mile south on the First Ordnance Survey map, 1813.
4. *Sussex Weekly Advertiser.*
5. Kelly.
6. H.E.S.S.
7. S.A.C., photograph by Mr. Beasley.

Alfriston, page 128

1. H.E.S.S., from Mr. E. Hewitt.
2. *Sussex Advertiser,* 19 June 1835.
3. Kelly.

Angmering New Mill, page 129

1. H.E.S.S., from the Grant family.
2. *Evening Argus,* 24 January 1973.

Crowborough, page 130

1. Gatsby, 1876.
2. Kelly.
3. H.E.S.S.

East Wittering, page 131

1. Information supplied the Redman family.
2. W.S.R.O., Tithe map, TD/W51.
3. Kelly.
4. H.E.S.S.

Eastbourne, page 132

1. Y. J. Lovell and Co.
2. Rev. P. Hemming, from Mr. W. J. T. Coppard.
3. Kelly.
4. H.E.S.S. from Mr. H. R. Hawksley.

Mark Cross, page 133

1. H.E.S.S.
2. Kelly.
3. 12 August 1856.
4. Rev. P. Hemming, *Windmills in Sussex,* 1936.
5. Notice in *The Miller,* 31 July 1911.

Worthing Navarino, page 135

1. Researched by H. L. C. Smail, directory references, etc.; also an article in *The Worthing Herald,* 28 January 1955.
2. Photograph taken in 1895, Worthing Reference Library.

Eastbourne Horizontal Mills, page 135

1. Mr. Lawrence Stevens, Archaeological report scheduled for S.A.C., Vol. 116.
2. A copy of a local map showing this may be seen at Polegate Windmill museum.
3. H.E.S.S.
4. This is reproduced in *Along the South Downs,* by David Harrison, Cassell, 1958, p. 46.

Ditchling Utility Mill, page 137

1. H.E.S S., from Mr. Hollis.

John Olliver, page 146

1. From an indenture of a lease in the possession of Mr. G. Frank.
2. G. V. Irving in the *Journal* of the British Archaeological Association, 1857, p. 289,
3. Notice in the *Sussex Weekly Advertiser,* 12 November 1827, shows sale of the mill gear.
4. Funeral account, *Sussex Weekly Advertiser,* 29 April 1793.

Cooper, page 149

1. Pigot.
2. W.S.R.O., Tithe map TD/W 157.

Holes, page 151

1. I am indebted to Mr. Edwin Hole for a full account of the history and working details of this family firm.

Holloway, page 153

1. Kelly.
2. F. W. Gregory.
3. Worthing Reference Library.
4. H.E.S.S., from the Harris family.

Medhurst, page 155

1. Pigot, Mr. Aubrey Ruff, a descendant, understood that Samuel first set up as millwright in 1820.
2. 'Sussex Millwrights', by Aubrey Ruff, *Sussex County Magazine,* 1952, p. 8.
3. H.E.S.S., from Alderman John Every.
4. Samuel was six in 1805, and William eight.

Neves, page 157

1. From Mr. Sidney Ashdown who knew the Neve family.
2. Melville.
3. Kelly spelt Steven in every case; however, his gravestone shows Stephen.
4. Kindly lent by Mr. S. Ashdown.
5. Still to be seen.

Obscure Remains, page 165

1. Based on H. E. S. Simmons' notes, Brighton Reference Library.
2. *West Sussex Gazette,* 29 October 1857.
3. Bennett and Elton, *History of Corn Milling,* 1899, p. 277.
4. Mr. F. W. Gregory.
5. Mr. E. Hole.
6. Mr. Tant.
7. My thanks to Mr. H. R. Hawksley for help here.
8. J. M. Baines collection, Hastings Reference Library.
9. Mr. J. Cornford.

Model Mills, page 169

1. Mr. R. J. De Little, author of *The Windmill Yesterday and Today*, John Baker, 1978.

Bishopstone, page 170

1. *The Engineer*, 12 July 1901, and 1st edition 6-inch O.S., 1873.
2. Mr. W. C. S. Tubb.
3. Newhaven and Seaford Historical Society.
4. Official heritage History, 1948

Birdham, page 170

1. H.E.S.S.

East Blatchington, page 172

1. H.E.S.S. from the Newhaven and Seaford Water Company.

Punnett's Town saw smock-mill, page 172

1. H.E.S.S., from the Lower family; also G. M. Fowell in 'Windmills in Sussex', *Walker's Quarterly*, Nos. 29 and 30, date scratched on a window pane, 27 June 1866.
2. H.E.S.S., from Mr. H. R. Hawksley.

Twineham, page 173

1. H.E.S.S.
2. From Mr. Edwin Hole.

West Ashling, page 173

1. Rex Wailes in *The English Windmill*, 1954, p. 83.
2. S. P. B. Mais in *England of Windmills*, 1933.
3. Rev. P. Hemmings in *Windmills in Sussex*, 1936, p. 98.
4. Mr. F. W. Gregory.
5. Silver Jubilee Historical Survey of Funtington, 1977.

Angmering, page 175

1. From a booklet, *Windmills*, by R. T. Hopkins, p. 23.
2. Rev. P. Hemming in *Windmills in Sussex*, 1936, p. 119.
3. S.A.C., photograph by Mr. Breasley.

Heathfield Wind-wheel, page 175

1. From Mr. Sidney Ashdown.

Punnett's Town Wind-wheel, page 175

1. Kelly.
2. H.E.S.S., from Neves.
3. Mr. T. Cornford.

Page 176

My thanks to Mr. H. R. Hawksley for helpful criticism here.

INDEX AND GLOSSARY

(*Figures in italics denote plate numbers; Roman numerals denote Fig. numbers*)